PENGUIN BUSINESS

A BIOGRAPHY OF INNOVATIONS

R. Gopalakrishnan has been a corporate leader for fifty years—the first phase of his career with Unilever lasted thirty-one years, while the second was for nineteen years as director of Tata Sons. Gopal spends the current—third—phase of his career writing, advising, speaking and teaching. He serves on select boards and advises companies. He is the bestselling author of several books, including *The Case of the Bonsai Manager*, *When the Penny Drops*, *What the CEO Really Wants from You*, *A Comma in a Sentence* and *Six Lenses*. Gopal is an international speaker and is actively engaged in both instructional and inspirational speaking.

He serves as a distinguished professor at IIT Kharagpur and as the executive-in-residence at the S.P. Jain Institute of Management and Research, Mumbai.

He can be reached at rgopal@themindworks.me.

T0182213

ALSO BY THE AUTHOR

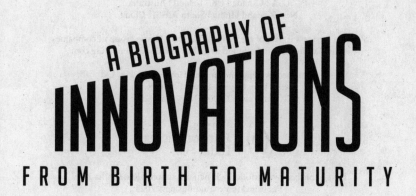

A BIOGRAPHY OF
INNOVATIONS
FROM BIRTH TO MATURITY

R. GOPALAKRISHNAN

BUSINESS

An imprint of Penguin Random House

PENGUIN BUSINESS

USA | Canada | UK | Ireland | Australia
New Zealand | India | South Africa | China

Penguin Business is part of the Penguin Random House group of companies
whose addresses can be found at global.penguinrandomhouse.com

Published by Penguin Random House India Pvt. Ltd
4th Floor, Capital Tower 1, MG Road,
Gurugram 122 002, Haryana, India

Penguin
Random House
India

First published in Portfolio by Penguin Random House India 2017
Published in Penguin Business 2021

10 9 8 7 6 5 4 3 2 1

ISBN 9780143456186

Typeset in Adobe Garamond Pro by Manipal Digital Systems, Manipal
Printed at Replika Press Pvt. Ltd, India

www.penguin.co.in

Contents

Contents

Preface

Innovation has long fascinated me. Even though I never applied my mind to the subject in any serious way, and this attraction was a more general one, in 2004 my relationship with the subject changed.

My company, Tata Sons Limited, requested me to chair a Tata Group Innovation Forum (TGIF). My steering committee had a cross-company membership, strengthened with subject experts. Ratan Tata's mandate to the TGIF was to positively influence the *culture* of innovation more than the *process* of innovation in all Tata companies. It was a challenging mandate as culture is a fuzzy subject to grapple with, especially for a whole bunch of engineers. Further, the effects of one's actions are difficult to ascertain as compared to technical measurements.

This mandate changed the terms of my engagement with the subject of innovation. I chaired the TGIF for eleven years. Naturally, I had access to all the Tata companies which operate across a wide variety of technologies, domains and consumer spaces. These companies gave me enormous insight about technology, business models, ambitions, human behaviour and persistence.

During this period I read a veritable library of books on the subject and spoke to international companies like DuPont, GE, Honeywell and DSM to learn and share. I participated in the workshops and symposiums of institutions like Imperial College London, London Business School, Stanford University, *The Economist* and the *Financial Times* as well as Indian institutions of distinction, like the National Innovation Foundation, the Marico

Innovation Foundation, the National Innovation Council and the Atal Innovation Mission.

The need to reflect and express my thoughts and observations was fulfilled in 2012 when I became a columnist for *Business Standard*. My monthly column, titled 'Innocolumn', was about innovation, entrepreneurship, creativity and allied themes. By the time this book is in print, I will have completed over fifty monthly contributions. All of these experiences have, in fact, inspired me to write this book.

A casual inquisitiveness about what happens in the human brain during the process of thinking, or the way in which it works, was an important part of my journey. If the reader finds the initial chapters somewhat rich with this layman's version of insights on neuroscience, my immense curiosity about the brain may be blamed.

It is this curiosity that led me to read about Michael Merzenich, who has done a great deal of work on the plasticity of the brain.[1] He says that the brain is plastic, in the sense that it assumes the form of the influences it is subjected to. The surface area of the brain has been likened to real estate by calling it 'cortical real estate'. I learnt that the brain does not waste any cortical real estate and any new learning occupies brain real estate. The more a learning is practised, the deeper that learning imprints itself on to the plastic brain.

This is a bit akin to how pathways get formed on mountain trails. New pathways don't come up easily, but the repeated use of any pathway strengthens it into an oft-used pathway. I learnt that 'in the five cups of tofu-like tissue inside the head, nestled amongst a trillion support cells, some 100 billion neurons are signalling to one another in a network with about half a quadrillion connections called synapses'.[2]

The learning itself may produce bad or good habits. There is a competitive plasticity in the brain to accommodate learnings of the individual. This also teaches us that not only is learning embedded in the brain, but that the brain itself is learning to learn! Amazing, isn't it? Thus, it is our habit and desire to learn, to be curious, that influences brain plasticity.

Providing me contemporary evidence was a report about John B. Goodenough. He set the technology industry ablaze at the age of ninety-four.[3] He and his team at the University of Texas, Austin, filed a patent application on a new kind of battery that 'if it works as promised, would be so cheap, lightweight and safe that it would revolutionize electric cars'. Thirty-seven years earlier, when he was fifty-seven years old, Goodenough had co-invented the lithium-ion battery.

My contemporary at the Indian Institute of Technology Kharagpur, Prof. Anjan Bose, who is a regents professor of power engineering at the School of Electrical Engineering and Computer Science at Washington State University, confirmed that age is no impediment to new learning, though motivation and determination can become impediments.

Bose, a great admirer of Goodenough, had the privilege of studying and advocating why the latter should be considered the father of lithium-ion batteries. Bose had served for several years on the committee that awards the Draper Prize, an American prize given through the United States National Academy of Engineering to technologies that have changed global society. The technologies are easy to identify and it is the committee's responsibility to identify the people behind them. Often, major technologies like the cell phone have dozens of key contributors. The Draper Prize, however, is given only when the key contributors are five or less.

In 2014, the Draper Prize was given to two academics and two corporate researchers—Goodenough, Rachid Yazami, Akira Yoshino and Yoshio Nishi. Goodenough was the first to make a Li-ion battery work in his lab and he is the only one still doing full-time research. During his fact-finding on the Li-ion battery, Bose spoke with many who knew Goodenough well. People were quite awed by the fact that even at ninety-four, he attends conferences, presents papers and collaborates with other researchers. He could very well get the Nobel Prize, says Bose.

This view contrasts with the Silicon Valley obsession with young people, epitomized by Mark Zuckerberg's comment at Stanford University in 2007: 'Young people are just smarter.' Author Pagan Kennedy writes, 'There is evidence to suggest that late blooming is no anomaly,' quoting a 2016 study that shows that 'inventors peak in their late 40s and tend to be productive in the latter half of their careers'. Nobel winners apparently make their discoveries at an average age of fifty.

It is important to note that age is not necessarily an impediment to being innovative, unless you have determined that to be so. What caught my attention was the evidence that from the time the innovator or scientist is born, it takes about forty to fifty years for him or her to peak and become truly productive. This goes against the generally held belief that for middle-aged people, great leaps of imagination are already well behind them.

I also found, anecdotally, that in science, the average time between a concept and its manifestation as a commercial product could be thirty to fifty years. The narratives in this book about climate control, the dry photocopier and the ballpoint pen adduce to this point. However, such a statement cannot be generalized across all products, domains and periods of time.

I started to see a connection between the life of a concept (and its later manifestation as an innovation) and the life of an embryo (and its later manifestation as a human being). Both seem to take forty to fifty years to peak. Although there is no science to support my view, I imagined that the way the brain conceives an idea may be metaphorically similar to the way a human womb conceives a baby—a chance unison of some energies. That is how my journey with this book took on the inflection.

In the corporate world, executives struggle with the I-word. The number of articles and books on the subject are so many that it becomes tiring to figure out how to apply them. Consultants and authors who have specialized in innovation management abound. The naturalness of creativity and innovation seems to be buried in the adipose tissue of management jargon. This state of affairs reminds me of modern American research that is quoted to persuade Indians that although we have to breathe to live, we really don't know how to breathe properly. The ancient Indian arts (or science) of yoga and meditation have taken centre stage as the remedies for this malaise, though in a modern avatar.

Humans should learn to innovate naturally, just like fish swim and flowers bloom naturally. For the large part, such things have to be internalized and self-taught—good life, health, happiness, positivity, creativity, human relationships, not to forget parenthood and marital relations. Just like how there is still so much to learn about the disciplined practice of breathing even though it comes naturally to humans, in the same way, there is so much to learn about the disciplined practice of innovation even though innovation is natural to human beings.

And that is my attempt in this book. I figured that a strong comparison of ideas and innovation to human life accompanied by 'stories' could provide distinctive lessons. Why stories? As averred by a behavioural psychologist, 'A story is about significant events and memorable moments, not about time passing . . .

The ending often defines its character . . . The remembering self composes stories and keeps them for future reference.'[4]

Changing the lens through which the subject of innovation is viewed helped me to see it in a much broader context. I picked up stories of innovation from anthropology (how humans evolved), archaeology (how Mohenjo-Daro was found), history (the last king of Burma), creative artistes (actor Shammi Kapoor) and religion (Jesuit training), apart from technology and products therefrom. I also talk about the crucial roles of psychology, sociology and human qualities. All this gave me a much broader view of innovation, which I felt impelled to share.

I'd like to thank the first reviewer of my early manuscript, Dr Sonal Jhaveri, former senior lecturer at MIT, Cambridge, and currently the science programme director at the Dana-Farber Cancer Institute, Boston. She patiently corrected my draft with respect to the facts of neuroscience, while I made my connections with life science and management innovation. Thanks also to my former Tata Sons Ltd colleagues Ravi Arora and Arup Basu for giving me their time generously. I thank Tata for the inspiration and insights that arose during my chairmanship of TGIF. I thank the newspapers where I wrote about business and innovation, particularly *Business Standard*, which gave me a platform to write my column.

And finally, my thanks to Milee Ashwarya and her superb team at Penguin Random House India, who enthused me with their contributions and infectious passion for the book. I have worked with Penguin before and the people there amaze the author in me with their skills and contributions.

R. Gopalakrishnan
1 September 2017
Mumbai

Introduction

The Life of Innovation in Eight Stages

I can imagine what is on your mind as you hold this book in your hands—'Oh no, not another book on innovation.' I cannot blame you. Innovation is among the most frequently used terms in management jargon of the last two decades. Everybody has a view on the subject and seems to write about it, and you wonder whether one listens to anybody else or reads everything that is written on the subject. And in the midst of this cornucopia of writings on innovation, here is yet another one.

Yes, you need to be persuaded that this book just might be different. And I will undertake this mammoth task of persuading you. I should start with the birth of the word 'innovation'.

A brief history of the word 'innovation'

Is innovation really increasing in the world? Is it the new buzzword? If so, is that good or bad? When the common man thinks of innovation, names like Apple, Google and the like come to mind. Why are there no such names from India's mighty corporations? Whatever cynics might say, buzz creates value as it changes language and culture, which is important in social transformation.

Tyler Cowen, an economist at George Mason University, Virginia, says that no matter what the hype may be, the march of technological progress in the world has hit a dry spell since 1970. 'Look at the period from 1870 to 1940 and think about how unbelievably creative and powerful that was,' he argues. He goes on to wonder whether a great ability to manipulate information has been exercised by a few people who keep talking about innovation. I get curious about the history and etymology of the word.

In Indian languages, there is no colloquial word for innovation. I emphasize on colloquial because I speak Tamil, Hindi and Bengali fluently. For words with a similar meaning, you have to resort to a Sanskritized artifice. Is that so in English as well? How and when did it come into English?

The etymology of the word innovation provides a clue.[1]

Canadian historian Benoît Godin states that the word 'innovation' first appeared in texts on law and meant 'the renewal of contracts'. It had no connection with creativity, and merely meant 'renewal'. During the seventeenth century, when Europe was far more conservative, every attempt to interpret religious matters was anathema. English puritan Henry Burton published pamphlets against 'church innovators' suggesting that such people were upstarts and imitators—not very complimentary. 'Innovation' in the early years carried a retrograde tonal quality.

With the Industrial Revolution, 'invention' became a highly desired activity, and by 1800, the word came into common English usage. Soon, this previously pejorative word started to get applied to science and entered the lexicon through a mental association with technical invention; perhaps this explains why, to this date, innovation is linked in many people's minds to research and development (R & D) rather than to business models.

This is quite unlike the case in emerging markets, where the term 'jugaad' has many contextual usages. An equivalent of jugaad exists in Brazil (*gambiarra*), in China (*zizhu chuangxin*) and in software ('kludge').

Google's useful Ngram Viewer, a database of word usage, suggests that in 1800, the word invention was used four times as frequently as innovation. After a century and a half, invention and innovation came to be equally used, and since 1970, innovation has overtaken invention and been used more frequently. The distance between the usages of innovation and invention is increasing. So what caused the pattern of word usage to change?

In 1939, for perhaps the first time, economist and political specialist Joseph Schumpeter differentiated between invention and innovation by stating that innovation involved commercialization and not just a new idea. From 1950 onwards, the American economy brimmed with growth and ideas. Innovation began to be thought of and funded as a packaged, predictable research product. Some experts like Cowen feel that innovation peaked around 1970.[2]

On 1 January 1969, a Bell Labs researcher, Jack Morton, and two scientifically trained journalists, William G. Maass and Robert Colborn, gave up their traditional professions to form the Technology Communication Company. This company placed advertisements in the *Scientific American* and the *Wall Street Journal*, inviting readers to become members of a new and elite group called the Innovation Group. The company also began to publish a magazine called *Innovation*.

A year later, in January 1970, 200 technology managers assembled for a workshop at Glen Cove, Long Island, to learn what it takes to be an innovator. They told stories about how their lives changed by the accelerating rush of innovation. Attendees were encouraged to seize the chance that was offered to them. This weekend workshop, for which participants paid the modern equivalent of $3000 each, created a club-like exclusivity with expert insight and collective self-help.

In hindsight, work by these three leaders created a buzz around innovation. Sadly, however, after the Long Island workshop, two tragedies occurred—first, Robert Colborn died of cancer and, second, Jack Morton was set ablaze in a gruesome murder. What a tragic end to two enthusiastic purveyors of innovation!

The human brain seems to have a design that does four things naturally: first, it solves problems; second, it solves problems relating to survival; third, it does so best in an unstable outdoor environment; and fourth, it does so nearly all the time, non-stop. The greater the survival challenges, the more frenetic the brain seems to get.

How the human brain has evolved

This part of the book may sound a bit dry. It is the layman's version to explain the human brain—where it began and how it came to be your most valuable possession. Unless you are a student of neuroscience, I think you will find this part enlightening, if not engaging!

Hundreds of thousands of years ago, our ancestors learned to walk on two legs instead of four. This change meant that they could walk longer distances in search of food.[3] Bipedalism also changed the role of the pelvis in the act of moving forward. A new analysis of the skull suggests that the evolution of the human brain may have been shaped by changes in the female reproductive system that occurred when our ancestors stood upright.

When our ancestors switched from walking on all four limbs to walking on two, it led to what is referred to as the 'obstetric dilemma'. The switch involved a major reconfiguration of the birth canal, which became significantly narrower because of a change in the structure of the pelvis.

Around the same time, however, the brain had begun to expand. In apes, the pelvis propelled the back legs forward, but for the two-legged, the pelvis became a load-bearing device, keeping the head above the grass. That act of walking on two legs freed up the hands and made the act of moving more energy-efficient. Our ancestors thus had a surplus of energy, which pumped up their brains. The quantum of blood that reached the cranium was different during our bipedal incarnation as compared to our quadruped equivalent.

Thus, biped walking changed the path and quantum of blood circulation in the body of our ancestors. The upright man survived and prospered for 2 million years.[4] About 300,000 years ago, our ancestors started to use fire, and the advent of cooking and the presence of light impacted what, how and when they ate. Teeth became smaller and the intestines shortened.

It was this sequence of events that caused the brain to grow disproportionately in size relative to the rest of the body. It seems we humans have a large brain in relation to our body structure as compared to other mammals. Further, humans developed a qualitative ability in what the neuroscience folks call the frontal cortex, through which we learned to think logically.

We also learned to express complex emotions, which other mammals find difficult to do. What is a complex emotion? It is the ability to not articulate literally what is on your mind. If an animal does not like you, it cannot express affection for you. Human beings can do so. So, the next time you warmly invite your neighbour home for a drink, intended as a social courtesy, both of you knowing full well that the invitation is not serious, you know that you have expressed a complex emotion.

The evolution of the brain is what caused humans to dominate the planet. Every other species adjusts to what it perceives as 'the realities', but the human is able to imagine a new reality, and work towards accomplishing that reality. About 10,000 years ago, our ancestors did not store food; they would grab a meal whenever they felt hungry and would eat whatever they could get their hands on. Their food was raw. They needed strong teeth to chew the skin or covering on the food item to get to its nutritious part. So, their jaws were stronger and their digestive system was designed to suit this particular way of accessing food.

At that time, some of them thought it would be a good idea to stop hunting from one meal to another. Why not settle down in one place where food could be cultivated? Why not domesticate some animals for their strength as well as their food value? This concept marked the beginning of the practice of farming, which had many innovative consequences. Humans learned which crops could be grown more easily, for which crops the yield could be improved, and how to store food to be eaten when hungry, rather than eat whatever is easily available.

Though this book is not about evolution and anthropology, it is pertinent to note that humans are distinctive in their ability to innovate. They have been innovating for centuries, and it is this ability that puts them ahead in their race against other species. It is through innovation that humans came to dominate the planet— regrettably, not always with positive, long-term effects.

The act of reproduction also went through evolutionary innovation. Most species have sex purely for the purpose of reproduction, so when a female chimpanzee comes into heat, she may mate with different males several times in quick succession within a short period, to improve the chances of conception. Thereafter, there may be a long hiatus in sexual activity. This is true for many mammals. The idea of sex for pleasure is distinctly human.

In many species, once the baby has been delivered from the womb, the mother's 'looking after' phase of the newborn is minimal. This is not so with the human species, and I always wondered why. This, too, was nature's innovation. With bipedalism, human brains grew larger while the pelvic structure and abdominal space reduced. In a quadruped, a foetus would stay in the womb for over nine months. However, the enlarged human brain of the baby made it difficult for the head of the newborn to emerge. So, humans started to deliver their babies a tad sooner, at nine months. In that sense, all humans are born 'prematurely', and the first three months after birth are sometimes referred to as the 'fourth trimester' of pregnancy. The human newborn requires protection and nurturing for a longer time as compared to most other species.

To further summarize this layman's version of anthropology, humans have always innovated, whether or not they knew the word. The human species itself represents innovations over thousands of years to become what it has today. Humans have a brain that functions quite differently from every other one of the 8.5 million species on the planet, and it has evolved through this long journey of innovation.

The executive network[5] is the part of the brain that makes us behave in a goal-oriented and focused manner. This alone, however, cannot achieve breakthroughs. There is another brain network called the creative network, which functions in a serendipitous and meandering manner. It is like a council of geniuses who discuss and exchange speculative theories and half-baked ideas. When the creative network and executive network function collaboratively, we move from concepts or ideas to innovations.

It sounds pretty quotidian to state the obvious—that humans evolved out of innovation, and to innovate is human! The convergence of genetics, information technology and neuroscience is casting a refreshing new light on the innovation process.

Metaphors and stories

We humans think and imagine through stories and metaphors. My personal insight arose from this world of storytelling and parables. It helps us make complex things simpler and assists us in dealing with what we don't readily understand in our day-to-day living.

We believe that a superpower runs this world, but for our daily living, we find it difficult to relate with this superpower. It is an idea, a concept. Most of us need something physical and tangible to be able to relate to an idea. We find it convenient to imagine a model that makes us comprehend a metaphysical concept. So we create and tell stories about God, we create images of God, and together, these make it easy for us to understand a complex world.

As humans became more civilized, they needed a code of conduct that would help run an organized society. This required us to define ideas like character, morality, courage and trust. How do we develop and communicate such an ethereal set of ideas? We tell stories! Our ancestors told stories to each other that resulted in wonderful books like the *Odyssey*, the Ramayana and the Mahabharata.

Over the centuries, different ways of telling stories developed and today we have a wide and diverse range of storytellers. India has an incredibly rich tradition of storytelling. The *bommai* tradition of Tamil Nadu, the ancient *anoiral* of the Meitei people of Manipur, the Mir musicians of Rajasthan, the *dodatta* of Karnataka, the *hari katha* in Tamil Nadu and the *jatra* tradition of Bengal—all of these add to the infinite variety of storytellers in India, indeed all over the world. A global example would be the storytellers of Jemaa el Fna in Marrakech.

Returning to the subject of innovation, would it help us to be 'natural innovators' if we could think of innovation as being quintessentially human and natural? Are we in danger of getting so trapped by jargon, techniques and methods that we forget to be creative and innovative? Fish don't take swimming classes, they just swim. Flowers are not taught to bloom, they just bloom. It is possible.

Innovation is natural and human. But we still need a way to think about the subject so that we can relate to it on a day-to-day basis. In this pursuit, metaphors and analogies can help us.

Can the process of ideation and innovation be related to the conception and development stages of human life? Do they have some similarities? Could the use of this metaphor, coupled with real stories about innovation, help provide a different view of innovation?

The eight stages of human life

Inspired by the stages of life enunciated by psychologist Erik H. Erikson, I attempted to consider the stages of innovation. I think it provides a reasonable and understandable model, probably because most of us raise children and relate to the stages of life.

Erikson is an ego psychologist. He proposed his theory of the many stages of life from infancy to adulthood.[6] During each stage, the individual experiences a crisis, which results in his or her psychosocial development. The incident or crisis may result in positive or negative development. The ego develops as it successfully resolves crises that are, for example, social in nature, like establishing a sense of trust in others, developing a sense of identity in society or helping the next generation to prepare for their future.

At each stage of life, the personality develops in a predetermined order because it builds on the previous stage. Erikson called this the epigamic principle.

Stating the research simply, these are the key features of the eight stages of life in the Erikson model:

The *first stage* is during the first few months of life after the birth of the baby. The child seeks a nurturing and loving environment through which it builds hope and trust for the future.

In the *second stage*, lasting up to the age of three, the child develops some skills, leading to a nascent but growing sense of independence and autonomy. The parents provide an environment of exploration for the child to try things, make mistakes and learn self-control without the loss of self-esteem.

In the *third stage*, which lasts approximately from age three till five, the child asserts itself through the vigour of its actions and behaviour, and demonstrates a sense of initiative, which the parents try not to curb.

The *fourth stage* lasts from age five till twelve, when the child learns to read and write and picks up skills that society is seen to value. The child shows pride in accomplishment and learns to build self-confidence.

In the *fifth stage*, between adolescence and eighteen years, the child searches for a personal identity. It experiences identity issues through the frustrating search for goals in its life.

The *sixth stage* is when this child becomes an adult and begins to show itself more intimately, and develop its own unique relationships.

The *seventh stage* is approximately from the age of forty till sixty-five, when the person settles down into relationships and starts to think of the bigger picture of life.

The *eighth* stage is what is referred to as a period of wisdom and maturity and occurs after the age of sixty-five.

This book is not about human psychology or the eight stages of life but about some rough equivalent of the eight stages of innovation. Inspired by the stages of life, I will tell you about the eight stages of innovation.

The eight stages of ideas and innovations

We can visualize the biography of ideas and innovations through eight life stages, which appear as eight chapters in this book. These stages do not bear literal parallels to biological phases.

The eight chapters are:

i.   How a concept is conceived in the brain (fertilization) and how this grows into an idea (pre-natal)
ii   How the idea is articulated through words or pictures (delivery)
iii The development of the idea into a demonstration prototype (infancy)
iv   The refinement of a prototype into a working model (childhood)
v   The product and business-model presentation (adolescence)
vi   The product competes in the market (adult)
vii The product achieves its potential (maturity)
viii The product renews itself to stay relevant to new innovations (ageing)

This book is not the biography of a single innovation but of innovations in general. This approach makes it possible to interestingly narrate multiple stories of innovation that illustrate the challenges of survival and growth in each stage. Hopefully, that makes for more interest for the reader and enhances the ability of the reader to relate to the social or corporate environment he or she is familiar with.

I have used several of these stories in my speeches, presentations and writings over the years, particularly in my monthly column, 'Innocolumn', in *Business Standard*, and my 'Ecosystem' column in the *Economic Times* on Sunday.

In the brain, a concept gets fertilized

Just as you need a male sperm to unite with a female egg for reproduction, you need neurons to fire together to fertilize a concept in the human brain. Just as you need millions of sperms to compete and one of them to unite with a single rare egg, you need many neurons to fire in symphony before successfully instilling a concept in the brain. A profuseness of brain neuron activity, randomness in the conception and nurturing care of the concept are the characteristics of this phase.

Concepts develop into ideas

Once a concept is lodged in the brain, it develops into an idea much like an embryo develops into a baby. However, a concept in the brain is about as interesting as an embryo in the womb. The magic happens when the originator of the concept is able to articulate his or her concept into an idea. This is similar to the way we recognize a child only when it is born and not when it is an embryo in the womb. Similar to the arrival of a newborn baby, the event of the idea being articulated is very important to the person but commonplace to others. The articulator of the idea may be blabbering with excitement, but the enthusiasm of and reception by others may not match his own zeal. As with human babies, this stage must be characterized by a loving and nurturing environment that builds hope and trust.

Ideas grow into prototypes

In this third stage, similar to a baby's infancy, an idea faces challenges to its survival. There will be toxins and viruses that may be threatening the baby's survival. The baby survives these threats and learns to develop its skills, leading to a growing sense of independence and autonomy. The parents in turn must allow the child to explore his or her ability by setting up an environment that is tolerant of failure.

The innovator's idea faces similar challenges. The peremptory dismissal of an idea, or discouraging any chance of failure, both deter the idea from developing into a prototype. The innovator learns self-control without the loss of self-esteem, so that the innovation can be built in a wholesome manner.

Prototypes adapt to become models

The innovator, like during a child's adolescence, gains in self-confidence in this phase. In order to develop the prototype, work on its shortcomings and produce a prototype that others can buy into, it requires the innovator to have a growing sense of autonomy and independence. Initiatives need to be taken to advance the innovation with good, interpersonal relations. The innovator, like the five- to fifteen-year-old, tries to balance initiatives with the guilt of taking things for granted.

Models are shaped as products for the market

This is the equivalent of children growing up from adolescence into adulthood, covering the ages of fifteen to twenty-five years. During these eventful years, they start to develop skills that are valued by society and look to their environment for encouragement to develop and demonstrate accomplishments. They also assert themselves as individuals with a personal identity. It is a period of some degree of confusion because the individual does not know exactly what he or she wants to be.

The innovator tries to arrange funding and sponsorship, but is faced with a barrage of challenging questions. Has consumer feedback been obtained? How is the prototype different from the competition? What is the basis for the sales revenue in the business plan? The innovator has no precise answers to such questions and feels a sense of irritation that there are so many to ask questions, but finds none to walk the path with him or her.

Products compete to grow

This stage of the innovation is a bit akin to the young adult aged twenty-five to forty years.

After passing all of these stages, the prototype has been produced for general use and is ready to be launched into the market. It is the debut of an adult entering the market space. Now its promise to the consumer, its delivery on that promise, its dissenters and its protagonists are all out there in the open to be publicly criticized or appreciated.

Apart from greater visibility, the product now begins to set up relationships with segments of the external and internal environment. Younger people like it, but not older people. The consumers of a particular ethnic or religious group relate to it, but not others. Internally, the production folks hate it as it impacts their productivity, but the logistics folks love it because it comes in easy-to-handle square boxes.

Products challenged to change

At this stage, the journey becomes easier to relate with the human life cycle as this phase is the middle-aged person, aged forty to sixty-five years approximately. The product has settled down into some relationships, seeks stability and develops a sense of the longer term. It may be the time when the product is vulnerable to newer wannabes, based on a more modern technology or presentation.

Struggle for relevance

This is the final stage of the product. It may be rejuvenated into a new avatar or may suffer silent attrition. Reflect on what is happening to fountain pens and wristwatches in today's age of digital handheld devices.

Nutrients at each stage

Like a growing baby requires different nutrition and care during each stage of life, innovation also requires nutrients and care. In this book, I have identified them to be somewhat as follows:

- BIRTH: serendipity—mindset—associative thinking—articulation
- INFANCY: open-mindedness—overcoming obstacles—risk-taking—continuous learning
- CHILDHOOD: learning from failure—adapting continuously—coping with opposition—developing the story
- ADOLESCENCE: disciplined development—acting fast but with thoroughness—getting lucky along the way
- YOUNG ADULT: process skills—social skills—networking skills

- MATURITY: observing beyond the obvious—sanitation, agriculture, hotels and cross-border skills
- AGEING: time from innovation to invention—application before the science—learning new things—innovation, entrepreneurship and change

The organization of this book

Each of the eight chapters in this book that correspond to the eight stages of innovation dwell on some features of each phase, supported by a number of stories. Hence, the writing is anecdotal rather than pedagogical. I believe that the reader would relate better to this approach as compared to a dry, theoretical dissertation.

I

Fertilization: A Concept Gets Fertilized in the Brain

Metaphors help us understand how concepts are generated in the brain. This chapter, therefore, contains simple ways to understand the human brain and its working. 'Simple', however, is never simple enough when it comes to the human brain. The reader may forgive the technicalities that may have crept in, but I do think it is useful to widely understand brain science.

Imagine a densely packed garden of flowers in full bloom, laden with pollen. A swarm of frenetic bees is at work, flitting around busily from flower to flower, transferring pollen (male) to the pistils (female), thus causing a riot of cross-pollination. In this way, the garden continues to bloom with an iridescent variety of flowers.

The garden of flowers can be used as a metaphor for reproduction in general. Think of the flowers as chromosomes, and the pollen as genes, or DNA, which encodes proteins. Each gene contains a unique code of life and the chromosomes carry these genes in 'full bloom'. Through the reproductive chaos (like the bees), the genes for a unique protein in a female will periodically encounter the gene for that unique protein in the male. Some genes miraculously mate to produce a new life that is neither one nor the other of the parent genes, nor a simple average of the two parents. This is how human and animal reproduction takes place.

The brain has 100 billion cells called neurons, which are communicating with each other incessantly through the generation and transmission of electrical charges. Here is approximately how it happens.

A concept in the brain is like an embryo in the womb

Creativity can be fostered, if not quite engineered—just like reproduction in animals and pollination in flowers! We routinely use the reproduction metaphor when we talk about ideas and innovations. For example, we use expressions like 'the innovation was born', 'how innovation was conceived', or 'how the idea was aborted'. How did this metaphor of an idea in the brain and an embryo in the womb come about?

Using this metaphor, it is possible to think of innovation as a human being with a life of its own—how ideas are conceived, how only some survive, how some are born, how they mature into infants and go through a life cycle. I find the metaphor greatly instructive and inspiring.

I set out to research whether this is just the poetic licence of writers or is there some scientific basis to it? If so, then an innovation must go through phases such as conception, nurturing, infancy, adolescence, adulthood and ageing, much like human life! It could provide unusual and creative insights into the innovation process—its reliance on chance, uncertainties in development, threats to survival and unexpected outcomes.

To recapitulate the human reproductive process, over 100 million sperms swim furiously for a very short time trying to reach one available female egg. Very occasionally, one of these sperms successfully unites with the female cell. The embryo is nurtured in the female womb and emerges as a baby in due course. Humans truly are products of the greatest lottery, we are one out of over 100 million other aspirants! Surely, this seems good enough reason to believe in luck!

As for the brain, it is an incredible piece of equipment about which we know rather little, though modern advances are dramatic and insightful. The brain cell, or neuron, has three parts or components.[1] First, there is a cell body that contains the DNA code and sustains the life of the cell. Second, there is a long wire, like the trunk of a tree, called axon, which carries the electrical signals. Axons are long, skinny and complicated 'things' in the brain—a bit like a stacked plate of noodles! Third and last, there are dendrites attached to the cell body, which appear like the branches of a tree. These carry the signals as far as they can and thereafter pass them on to the receiver dendrites.

The 100 billion neurons in our brain are in a constant state of apparent chaotic activity, flashing electrical signals through a complex wiring. When the neurons transmit the charge to the ends of their axon cable, they encounter an electrical gap. Contiguous dendrites are not touching each other, so how does a dendrite 'pass on' the electrical signal?

Fascinatingly, there is a real gap, called 'synapse,' between dendrites, and the signal is actually passed on across this gap. But the electrical signal does not travel wirelessly like a Wi-Fi or Bluetooth signal. Wonder of wonders, the information in the signal is converted into chemical neurotransmitters, which resemble sacs, or pouches of signal-encoded chemicals! The synapse is a thousand times thinner than the width of one human hair!

Artist, photographer, doctor and the father of modern neuroscience, Santiago Ramón y Cajal, won the Nobel Prize in 1906 for his neuron doctrine.[2] Long before modern devices became available in the early 1900s, Santiago made beautiful sketches of the brain, which have now become the subject of a travelling exhibition.

Like in the case of the bees, the neurons need something to get connected to. Nature provides this through neurotransmitters called dopamine and serotonin. When the electrical signal reaches the synapse, it releases neurotransmitters in the form of a chemical soup that floats across to an eager dendrite on the adjacent neuron.

For the most part, the neurons seem to communicate in a chaotic manner. Sometimes, a bunch of neurons begin to act synchronously, a bit like instruments in a chaotic orchestra suddenly starting to play in harmony. This is a magical moment, when the neuron activity results in a concept in the brain. But a concept is no more an idea, just as a zygote in the womb is no more a baby. A concept may have formed in the brain but a new idea has not been born. Scientists have found that the more chaotic the neuron activity in the brain, the higher the IQ of the person!

Neurons get activated by diversity, and by a change in routine activities and thoughts. Wrestling with an opinion different from your own, bending the mind to see an opposite view, going away for a holiday, taking a long walk, reading a new book, listening to music or even sleeping over a problem—all these get the neurons firing to provide new combinations of thoughts.

This licentious description, by combining known science with imagination and wordsmithing, helps a layperson understand how the brain works and how frenetic the activity is. Billions of neurons are thus firing all the time, discharging clouds of dopamine and serotonin in an uncoordinated and wasteful way—a bit like the disorderly traffic in cities like Mumbai or Bengaluru. It is an inefficient process of throwing away and wasting a lot so as to accomplish a chance encounter.

Occasionally, quite magically and mysteriously, these random neurons get highly organized and fire in harmony. Just like how in an orchestra, chaotic noise suddenly becomes orderly music, so also an idea emerges out of the firing neurons; and like how the zygote has several survival challenges, the conceived idea too is faced with survival challenges.

As a student of innovation, I draw several lessons from this metaphor. Firstly, that the formation of a concept in the brain is short-lived, unpredictable, random and wastefully inefficient Secondly, that just as human birth is a lottery of one in several million, so too is a concept in the brain. There are 100 billion energetic neurons in the brains of all the people in the world, but very few materialize into some form of reality. Having an idea is a random and statistical process. Thirdly, that the creative process is about as wasteful and inefficient as the reproductive system. And lastly, that just as there arise instant threats to the survival of the embryo, so it is the case with ideas too.

How concepts develop into ideas

Writer Steven Johnson brings considerable wisdom to the subject of innovation.[3] He has argued that brain concepts do not become ideas in the magical flash that storytelling suggests. How the apple fell on Isaac Newton's head, or what scientist Friedrich August von Kekule said when he got the idea of the benzene ring: 'Larger structures, turning and twisting in a snake-like motion . . . as if by a flash of lightning, I awoke . . .' The stories come often after thoughts. The concept lingers on, lurking in a corner of the brain, refusing to go away. New influences, other people's ideas and work on the concept (which never remains in its original form) are all similar to the ever-changing foetus in the womb. Johnson describes how the coffee houses of Europe gave a fillip to advance the ideas of the Enlightenment—an intellectual and philosophical movement that dominated the world in the mid-seventeenth century. Before this period, people just went to bars when they had spare time and money, because the quality of drinking water was suspect and alcohol might have been perceived as bacteriologically safer; its side effects must have also been pleasurable. The advent of tea and coffee promoted the concept of English coffee houses across Europe in cities like Budapest and Vienna, where people with different ideas and views congregated, and concepts that were lurking in their brains found rapid transformation into ideas that they could articulate.

Sometimes ideas flower, but for every idea that flowers, many more fall by the wayside. Timothy Prestero wanted to design the world's most perfect and affordable incubator for newborn children. It was a marvellous concept that would change the lives of the families of 4 million babies who were dying all over the world. His organization, Design that Matters, designed the concept into an idea, but it failed to go into production. In his YouTube video, he explains why the concept failed to go into the idea and innovation stages.[4] His company's concept won *TIME* magazine's award for the '50 Best Ideas in 2010', and his TED Talk in 2010, which explains why the concept failed, has been seen by over 1 million people.

Here are two stories of concepts that grew into ideas and innovations.

A natural way to fairness

I delved into my memory for an example I had personally witnessed, of fostered creativity, rather than rely on inspiring stories I had heard about from external sources. And here is a story from my career at Hindustan Lever Limited (HLL, now known as Hindustan Unilever).

The concept of becoming fair-skinned is not new to humankind, and even as this concept is embedded in the minds of several people, it develops differently into ideas, with each one leading to a different outcome.

The attitude towards skin colour varies with different countries and cultures. There are fair people who want a tan, and vice versa. Biologically speaking, the colour of the skin evolved as a response to the intensity of sunlight a person was exposed to. The desire for light skin resulted in the discovery of several formulations, including the bleaching of the skin—a harsh skin treatment that produces complications with continued usage. This bothered the biologist Girish Mathur. Who was Girish Mathur, you would ask?

The Hindustan Lever Research Centre began in 1958 in a spare room at the Bombay factory. The brilliant Cambridge-educated S. Varadarajan got together a bunch of highly qualified scientists with varied temperaments, backgrounds and disciplines. They were, however, all mentally prepared by the company's leaders to be passionately united in the quest for products that could advance business through the application of science and technology to everyday consumer products. It was a leap of corporate faith, rare thinking for the 1950s.

Mathur was one of the scientists recruited by the centre. In 1966, as a result of his studies on dermatology and skin cancer, Mathur became intrigued by what causes fair skin. His team delved deep into what science said about one's complexion. A lot of it had to do with evolution because the amount of ultraviolet radiation penetrating the skin has to be regulated. An important factor was a substance called melanin, and its distribution under the skin, which has a strong influence on complexion—a bit like how the quantum and spread of butter on a piece of toast influences its taste perception. Should the research team develop a product to regulate the distribution of melanin? They produced a prototype by 1971, but it sat on the shelf.

In 1973, the marketers in the office observed a very old Indian phenomenon: that the newspaper matrimonial columns were full of advertisements for 'fair brides'. They wondered whether it was possible to produce a cream that could make women fair naturally and without bleaching the skin. It was just the opposite of the Western practice of tanning the skin! This particular need was uniquely Indian, and if the Indian company did not work on it, nobody else would. How could a European even think of working on a fairness cream?

These thoughts aligned the minds of the scientists, the marketers and the production folks. It took two years of colliding viewpoints and frenetic neurons to formulate a safe product. The professionals argued fiercely: there were huge differences of perspective among the scientists and also between the scientists and the marketers.

Unilever in India had a chairman at that time, T. Thomas, who hailed from Kerala, where there are dark-skinned people. He took the entrepreneurial decision to go ahead with a test market in the south. As I now recall the events after forty years, the developments had a denouement with twists and turns, hope and despair. And the neurons kept communicating relentlessly.

Today, Fair & Lovely counts as an Indian contribution to Unilever. The brand has a revenue of about 1 billion dollars globally, not too far away from another global skincare brand called Dove. Unilever has the world's most successful natural fairness cream, with millions of loyal customers for Fair & Lovely, not just in India, but in other markets too where brown-skinned people desire to become fairer.

Keeping cool

Consider a time about a century ago when one lived and worked in either a hot place or a cold place. You could not change the weather so you equipped yourself for the weather. But, hang on!

Who thought up the concept that you could influence the weather around yourself? Whose idea was it, and how did it become an innovative product of its time? Has it been an evolutionary process over centuries?

Humans have long tried to improve the conditions they encountered naturally. In ancient Rome, the wealthy used an aqueduct system to circulate cool water through pipes embedded in their walls. In the third century, Roman emperor Elagabalus dispatched a thousand slaves to the mountains to fetch snow to cool his gardens. The hand fan has been in use for centuries, from as early as 3000 BC in China. Electric fans appeared in the US only in the twentieth century. In 1758, Benjamin Franklin and John Hadley discovered that the evaporation of certain liquids such as alcohol produced a cooling effect. Fifty years later, Michael Faraday found that liquefying ammonia could chill the air. In 1830, John Gorrie patented an ice-making machine over which he blew air to cool the air.

All of these, and perhaps more, were progressively improving innovation and taking further the concept of influencing weather.

In the early 1900s, a publisher in Brooklyn called Sackett-Wilhelms Lithographing and Publishing Company faced a scheduling problem. The company was printing a magazine and had to pass the paper through the printer repeatedly, once for every colour. If the weather was humid, the ink from the previous pass would not dry in time, thus delaying the next pass of the paper. This affected schedules and deadlines.

The company referred the problem to the Buffalo Forge Company. It was not as though this company had established expertise in this problem or its solution. The company did, however, encourage curiosity in people. Its founder, William Wendt, had caved in to the demands of a young and restless

Willis Haviland Carrier to create a laboratory to tackle somewhat speculative projects—like dehumidifying air.

A young engineering graduate from Cornell University, Willis had joined Buffalo Forge in 1901 at the age of twenty-five. As he thought about the air one foggy morning on a Pittsburgh train platform, he realized that he could influence the moisture in the air by passing dry air through water to create a mist. He was unaware that he was all set to serendipitously solve the problem posed by Sackett-Wilhelms.

This laboratory was the perfect location, and Willis was the perfect person to whom this problem could be referred to. The young team adopted a couple of approaches, which failed, and Willis used his instinct to devise an alternate solution. In the process, he invented modern air-conditioning. On 17 July 1902, his newfangled humidity-controlling machine began operations. The young innovator patented his invention and went on to create the Carrier Corporation, which grew to be a world leader in air-conditioning.

The story about electric bulbs, believe it or not, bears some similarity.[5] Historians Robert D. Friedel, Paul Israel and Bernard S. Finn have compiled the names of twenty-two people who invented the electric bulb before Thomas Edison even filed the patent for his product. One of these twenty-two was one John W. Starr, who got his patent in 1845, long before Edison. According to another book, Edison's first patent application was rejected because it infringed Starr's patented design from about 140 years earlier.[6]

If you were to write about the concept of influencing weather conditions or the idea of electric lighting, when would you say the concept was birthed? Whose neurons were responsible, and who is the real father of these innovations? Like much of history, it depends on who writes it.

Every idea and innovation seems to merely be an evolutionary punctuation mark—each with a past and a future. If society was more concerned with the number of people who would actually benefit from innovation rather than who the original inventor was, there would be an explosion in the innovative output all around.

But this story is not about Fair & Lovely or air conditioning and other innovations that are taken for granted these days by consumers. It is about what happens to a concept while it is 'trapped' in the brain and developing to be born as an idea.

An idea is 'born'

Any layperson knows that the embryo is the first step in reproduction, and after being in the womb for nine months, it transforms into a baby. However, it is up to scientists and people in the medical field to understand exactly how the embryo becomes a baby. With the power of visualization technologies, algorithms and deep science, esoteric subjects like these are explained in amazing terms through books and other digital media.[7]

I think that something equally spectacular occurs when a concept in the brain is developing to be expressed as an idea. As neuroscience advances and our understanding of how the brain works improves, we are bound to learn a lot more in the future.

However, it is common knowledge that prolific ideation is a great aid to creativity. Being prolific in generating concepts and solutions to problems actually increases originality, according to experiments by Northwestern University psychologists.[8] The research found that after generating the first twenty solutions to a given problem, people actually came up with more solutions. Fecundity is a virtue in the creative process.

Further, promoting dissent and alternative points of view spurred the fecundity of solutions. This catalyst of dissent as a trigger for prolific solutions is easy to understand but rather difficult to implement in organizations.

Both the stories of Fair & Lovely and of the invention of air conditioning demonstrate that a concept in someone's brain is of limited interest. The conversion of that concept into an idea through articulation gives the concept life. Most ideas have evolved over long periods of time through multiple concepts in people's brains.

Who knows whose concept was the original one or even what the original concept was? The undisputed original thinker was Adam in the Garden of Eden. Since Adam, everyone has built on others' ideas. Innovation, recognized as such, appears periodically when it is perceived as a novel, problem-solving version, even a rearrangement, combination, breakdown or mutation of existing ideas. The distinctive packaging and delivery of the solution to any problem is innovation, almost surely a build-up from a bunch of concepts.

That is why, in judging innovation and innovativeness, the impact on human lives should be thought of as being more relevant and important than the person who thought of the idea, or even the idea, per se.

On the contrary, everyone seems to be more obsessed with the person who first got a particular idea. Who has the patent? Who made a million or billion bucks out of the idea? What do we know about that person? Books and articles love to focus on the lone creative mind, and that person then becomes the hero: how Thomas Edison invented the electric bulb, Willis Carrier the air conditioner or Vic Mills the disposable diaper!

Writers on innovation praise the person and dramatize the circumstances under which his or her mental light-bulb flashed.

Like how success has many fathers, well-known innovations, too, have many fathers. Unlike human beings, for whom paternity can be clearly established through DNA testing, paternity in matters of innovation is complex to establish!

In a nutshell

Much drama is associated with how a concept strikes root in the brain. It has been eulogized as being a flash, an epiphany or a sign of genius. The reality, however, is that neurons are active in the brain all the time. For most of time, their signals are not in symphony, but once every so often, through what appears to be a random process, there is a coming together of the chaotic signals in the brain. This can be thought to be the formation of a concept. However, a concept is as interesting and valuable as a foetus. Much remains for the concept to get delivered into the world as an idea. That is the subject of the next chapter.

II

Birth: An Idea Is Born Out of This Concept

Whenever I have likened the fertilization of a concept in the brain to the development of an embryo in the womb, my technically sound friends would look at me with an 'Are you sure?' look.

Author and science journalist Matt Ridley has delivered a very interesting view to support this idea.[1] He has argued well the view that throughout history, the engine of human progress has been the 'meeting and mating of ideas to make new ideas'. It is not important how clever individuals are, what really matters is how smart the collective brain is!

In this chapter, I trace how a concept develops into an articulated idea over time. Just like how an embryo derives all its nourishment from the mother, a concept too gets all of its developmental nutrients from the brain of the person in whose head it resides. Just as the muscles and hormones of the mother undergo miraculous transformations while the embryo develops into a baby, the conversion of a concept into an articulated idea causes astounding tensions, paranoia and dreams in the brain of the person with the concept. That is why I use the metaphor of life for the biography of a concept.

There are four active promoters that help develop a concept into an idea. They are:

- Serendipity
- Mindset

- Associative thinking
- Articulation

Serendipity

I've always wondered where the word 'serendipity' originated from. The Persian tale *Three Princes of Serendip* inspired art historian and the fourth Earl of Oxford, Horace Walpole, to use the word in a letter in the 1870s. Serendipity, a word derived from the Arabic language, is the occurrence of events by chance that produce beneficial results. It is a crucial word for innovation. The problem is that we do not recognize the moment of serendipity, neither are we capable of planning it. We take it as it comes.

Thus, ideas arise out of frenetic neurons, but how do these neurons become frenetic? How does serendipity work? What promotes serendipity? I owe some of this inspiration to already published ideas and thoughts.[2]

Dreams arise as a result of the mind experimenting with new combinations of neurons. Dreams are, therefore, an act of exploration of new things by the mind, rather than Sigmund Freud's concept of dreams unveiling a repressed truth! The concept, however, has to develop just like the foetus does, before becoming a baby. The equivalent of the baby being born is the ability of the conceiver to articulate the idea through words or drawings.

Articulating democracy, flying machines and the benzene structure

How, for example, would you explain what democracy and adult franchise are to people who have no words for it and, hence, no conception in their mind of these terms?[3] A communications professional explains what it takes to explain 'elections and democracy' to those in Afghanistan and South Sudan who just don't know these words. She references the Aristotelian tautology: 'If something does not exist, there is no word for it; and if there is no word for something, then it does not exist.' It is fascinating to listen to her explanation of how her organization explains elections and democracy to people who have no knowledge it and, therefore, have never experienced anything like it. She explains the methods and tools used by her to present a concept as an idea. Democracy is well understood by us, but is to be explained to such people through an articulated idea, so that they understand it.

Long before science knew about flying machines, the concept of flying chariots was mentioned in India's Rig Veda and interpreted to mean 'mechanical bird'. It is an inspiring tribute to the history of ideas, that the concept of something that is heavier than air could be thought to fly at least twenty-five centuries before the Wright brothers made an aeroplane. The sage Valmiki imagined a flying machine in the form of the *pushpaka vimana,* which was described as '[resembling] the Sun, and could go everywhere at will . . . Resembling a bright cloud in the sky . . . could rise up into the higher atmosphere . . .'

Most of us would associate Leonardo da Vinci with paintings like *Mona Lisa* and *The Last Supper*. He sketched ideas in notepads that have been preserved since his death in 1519. For example, da Vinci drew flying machines at a time when the idea did not exist. He would buy birds in the market, study their wing structure and shape and then set them free. He created one of his drawings by observing a small, agile kite and converting its skeletal structure into a flying machine. One of his sketches has a glider-like drawing with outstretched bat wings, and a noting of its proportions. There are details in his sketches: for example, in one of the devices he has drawn, there is a place where a man would lie down into the machine and his waist would be in a ring just below the wings.

Leonardo da Vinci did not have access to aluminium and synthetic cloth, but historians have tried to assemble the design he prepared with modern materials and a few modifications. They found that the resultant machine could indeed fly! Obviously we don't have scan images of what went on in da Vinci's brain. It is, however, entirely reasonable to assume that when he saw these birds in the market, his neurons were firing, and though he developed a concept, he could not articulate it. The concept sat in his brain as he struggled to articulate the unimaginable—that man could make a flying machine. His close observation of the birds in the market helped him draw out his concept, and it was this drawing that became the 'baby in the maternity ward'. It took, however, four centuries for that 'baby' to grow up into an 'individual', when the Wright brothers developed the first flying machine in the twentieth century.

The story of how German scientist Otto Loewi dreamt of the neurotransmitter also illustrates the conversion of a brain concept into an articulated idea.[4] Back in 1921, it was not known how neurons or brain cells talked to each other. Loewi was interested in learning how the brain signals the heart to speed up or slow down, depending on the circumstances. A concept thus developed in Loewi's brain—that a long nerve (called vagus), emanating from the brainstem and attached to the heart, exudes some liquid (*vagustoff*) that sends instructions to the heart.

His problem, however, was that he could not articulate the concept. So he connected two frog hearts, and when he subjected the vagus nerve to an electric shock, the heart slowed down. His hypothesis was the same—some liquid oozed out of the brainstem whenever required. He slept over the matter. One night, he woke up with an insight on how to do the experiment. Unfortunately, the next morning, he could recollect little of what had appeared clearly to him the night before. On three occasions, this charade of insight followed by the loss of memory continued. Finally, he did the experiment on the very night he had the dream. He could thus prove the existence of the liquid (acetylcholine) that comes out of the vagus nerve. Loewi got the Nobel Prize for his work. Now, science recognizes that acetylcholine is one of the neurotransmitters that neurons use to communicate with each other.

That is why, with regard to the creative process, the life cycle of an innovation begins as a concept in the brain, which manifests only when it develops into an articulated idea.

Controlling human reproduction

Mammals usually mate with the clear goal of reproduction. They do not mate for pleasure as far as we know. We *Homo sapiens* mate for pleasure and not necessarily for reproduction.

Hence, the quest for birth control is absent among animals, but ancient among humans, who have toyed with the concept from as early as the Egyptian and Vedic times. Until the twentieth century, birth control was about physically preventing the fertilization of the egg and sperm through use of materials such as diaphragms. Public discussion about sex and its practice were taboo because social attitudes and religious beliefs discouraged openness.

That was the case for centuries until 1945. At about the time I was born, quite coincidentally, the subject of sex burst forth into the open and the concept of birth control developed into a chemistry-based solution, rather than being engaged in the only known physical solution of blocking fertilization.

In 1937 three professors at the University of Pennsylvania discovered a 'chemical switch' during their scientific work with female rabbits. They published a paper that stated that if a hormone called progesterone is introduced into the female body, the female would actually stop producing eggs. And without producing her egg, she could not become pregnant! Bingo! It was that simple.

A revolutionary social idea developed simultaneously when a college professor in Indiana, Alfred Charles Kinsey, published his research findings stating that human beings, both male and female, were 'friskier' than they cared to admit. A graduate student at Northwestern University, Hugh Hefner, read the Kinsey findings and set out to rid America of the darkness and taboos concerning sex. Hefner went on to found *Playboy* magazine. This combination of chemistry and sociology set the ground to completely change ideas regarding human birth control.

Beginning in 1937, when the concept of progesterone was fertilized in the brain of scientists, through the 1950s, when the concept was manifested as a pill idea, and finally into the 1960s, the world approved the first contraceptive pill for marketing. This sequence illustrates very well the metaphorical stages of conception, birth and innovation.

Other events around the 1940s and 1950s also shaped and influenced the crystallization of ideas. A feisty American lady called Margaret Sanger had, for a few decades, been promoting birth control by starting the first birth-control clinic in 1916. She wanted to liberate women from the uncontrolled cycle of unwanted pregnancy by popularizing safer and more reliable methods of birth control, rather than relying on clumsy diaphragms and time-related abstinence from sex. She wanted the thirty states in America to revoke their anti-birth-control laws so that innovation in birth control could be encouraged. Margaret Sanger was considered dangerous by a society that was fearful of the consequences in case she became successful in liberating women. What would happen to family, marriage and moral standards?

On a cold winter day in 1950, sitting in her Park Avenue apartment, Margaret Sanger met one Gregory G. Pincus, an out-of-the-box-thinking scientist, and made her pitch to him. Pincus had already earned the reputation of being a Frankenstein-like scientist due to his experiments with the fertilization of rabbits. Sanger asked Pincus to make a pill that a woman could consume as freely as brushing her teeth, thereby allowing her to lead an enjoyable life without worrying about the risk of pregnancy.

Pincus had already developed the chemistry concept into an idea, but he had to experiment and demonstrate what he had thought up. In short, Pincus had a solution for a problem that nobody had posed. Of course, it was possible, he said. He needed $3100 to engage an assistant and buy some materials. Sanger could offer him $2000.

Unbeknownst to both Sanger and Pincus, another scientist in Mexico, Carl Djerassi, had already synthesized a hormone pill from wild yams, which could prevent ovulation in women. By 1960, the U.S. Food and Drug Administration approved Enovid, a contraceptive pill released by G.D. Searle, LLC.

Time magazine reported in 1966: 'No previous medical phenomenon has ever quite matched the headlong US rush to use oral contraceptives, now universally known as "the pills".' The concept of birth control has had a hazardous biography over centuries, but the pill has elegantly (and controversially) emerged as a new idea in birth control, arguably qualifying as one of the biggest innovations in human history.

Mindset

Through her research, Prof. Carol Dweck of Stanford University suggested that there are people with two types of mindsets: one, 'fixed mindset' people, who believe that their abilities are fixed and repeatedly use those abilities to carefully avoid failures; and two, 'growth mindset' people, who believe that their abilities can be expanded and accept that this involves learning from mistakes and failures.[5]

Young people are thought to have a 'growth mindset', which is likely to be experimental, while the old are thought to be of a 'fixed mindset', which is likely to be more risk-averse. Of course, every person sometimes has both mindsets, with some people having a greater inclination to either of the two. There are always exceptions, though.

Those with a 'growth mindset' are prepared to fail, but they learn from their experiences. The story of Indian film actor Shammi Kapoor illustrates how he displayed a 'growth mindset' by reinventing himself time and again.

Experimental actor Shammi Kapoor

The autobiographies of actors from the Indian film industry often discuss their failures more than their successes—quite contrary to the autobiographies of bureaucrats and management leaders! Born in 1931 as Shamsher Raj Kapoor, Shammi was a scion of the eponymous Kapoor film family. Being the son of Prithviraj Kapoor and the younger brother of Raj Kapoor put a lot of pressure on this upcoming actor. Shammi, however, remained an experimenter and learner throughout his life. Because he came from an illustrious film family, people expected that Shammi too would be yet another superstar. He, however, cut a different mould for himself.

Many of his early films were flops. He learnt to become a distinctive actor through his improvisation and dancing. He had his successes in the 1950s and his career peaked in the 1960s. Yet, he yearned for something different. In trying to carve out a distinct space for himself, he tried his hand at producing films like *Manoranjan* and *Bundal Baaz*. Both failed at the box office, but were acclaimed by critics as being ahead of their time.

Next, he dabbled with a music radio station that was again unsuccessful. Finally, in the early 1990s, he embraced the

Internet—something that was least expected from a dancing, light-hearted actor from the Indian film industry.

Kapoor seemed to innovate by experimenting, learning to do new things and enjoying himself. He did not seem to be trying to prove anything. He remained an innovative trendsetter throughout his career and displayed a 'growth mindset'.

An individual like Kapoor could do such things, but how do you inculcate this mindset into a large business or a whole institution?

Company examples

The experience of a chemical company called Rallis India gives us that example. The company was producing 20,000 litres of liquid waste, or effluent, from a manufacturing process. The company was not permitted to destroy this waste except through a common incinerator installed in the chemical zone. This brought with it the penalties of sustainability and cost overloads.

Could the company achieve zero-effluent discharge?

On finding that there were three effluent streams that mixed together to form the discharge, the company innovators asked, 'Why not deal with each effluent stream as it gets generated, rather than all the three together at the end?' Through an innovation programme, alternatives were attempted. Recycled water and useful organic salts were recovered and the effluent load was reduced from 20,000 litres to only 2000 litres per day.

The project has still not succeeded in achieving its zero-effluent goal. But the company nurtured a 'growth mindset' among its executives by encouraging honest and disciplined experimentation and daring its executives to give it a try.

I hesitate to narrate the whole story here, but Hindustan Lever changed the mindset of its managers by setting up a Research Centre in 1958 and encouraging its managers to think out of the box. If India was short of vegetable oils for making soap, what forest oils could be tapped and how could the company make soap out of exotic oils like sal, neem and karanja? Nobody had done so before. Well, Hindustan Lever did! Over the course of several years, the company pioneered the use of forest-seed vegetable oils for making soap during the 1970s—the first such feat in the world at the time. Above all, it saved the company from extinction due to a lack of traditional raw materials!

Tata Steel found a solution to counting its reinforcing bars or rebars through its software company, Tata Consultancy Services (TCS). A novel method to manufacture a consumer water filter was established by Tata Chemicals from the discrete manufacturing expertise of its wristwatch company, Titan.

Kingsford Coal

Unfortunately, companies often tend to adopt a 'not invented here' syndrome.[6] Very often, best practices can be found in completely different industries. Here is an example of the outcome of a knowledge-sharing trip, when a DuPont titanium dioxide team visited an obscure coal company.[7]

The DuPont executives manufactured a white powder and wore formal, button-down clothes. The Kingsford Coal executives made a black powder and wore sweatshirts and red suspenders.

The DuPont executives found nothing to learn from their visit and expressed their dissatisfaction to the consultant who had organized the trip. The Kingsford Coal executives, on the other hand, were enamoured by DuPont's safety standards and posed lots of questions to their visitors. They also took detailed notes.

A year later, while the DuPont executives had forgotten the visit, the Kingsford Coal executives had implemented the safety lessons and sought a meeting to thank DuPont. The Kingsford

Coal people displayed the 'growth mindset', while, in this case, the DuPont executives demonstrated the 'fixed mindset'.

Leadership in an institution, company or laboratory is a formal position. The mere act of achieving an exalted status can cause a leader to give up the growth mindset. Stakeholders and the media love the evergreen idea of an always-successful leader with a blemish-free track record. Why would such a person take a risk that can result in failure or bring criticism?

The reality is that holding a leadership position is like having a licence to drive. It does not mean that you will indeed drive skilfully under severely challenging situations. To do that, you have to take risks and experiment.

Leadership means creating for employees a meaning in the work they do and fostering engagement in their hearts. A 'growth mindset' allows a person to love what he or she is doing. They are not necessarily trying to get to the top rung, but they enjoy what they do so much that some of them do end up in the top rung. It encourages people to value whatever they are doing, irrespective of the outcome.

Associative thinking

Associative thinking is unusual in so far as it ascribes a value to inexperience. Experience teaches which solution has worked in one context, whereas inexperience encourages trying that solution in a different context. Interesting outcomes arise at the confluence of experience and inexperience.

- Robert W. Kearns, the inventor of the variable-speed windshield wiper, was driving his car on a rainy day and wondered why the wiper cannot change its speed with the intensity of rain, just as the human eyelid can increase or decrease the number of time it blinks.

- Vaseline was created by Robert A. Chesebrough, a chemist, who discovered that the gooey stuff protects human skin by sealing it, just as fog traps moisture on the ground.
- The Tata Swach water purifier removes bacteria and viruses by using a natural substrate (no chlorine, no UV rays) inspired by the way nature produces pure groundwater.
- The McLaren 12C supercar emulated the way water is channelled around the fins of a sailfish.

Such innovations exemplify associative thinking, and promoting this in an institution is challenging because it requires having a genuine respect for inexperience.

This runs contrary to our mindset that places experience on a pedestal. An acclaimed leader is often recognized for what educationist Howard Gardner called the 'disciplining' mind, which is highly focused on knowledge and systems to exploit resources efficiently. On the other hand, innovation thrives when the mindset leans towards exploration and imagination.

From experience, one can state that associative thinking can be encouraged in several ways.

The *first* way is for leaders to ask 'charging-up' questions that pose a ridiculous challenge.

A charging-up question provokes a response in the form of innovation. As a nation, India asked this kind of a question in the 1980s, when America denied India the Cray supercomputer technology. Consequently, Vijay Bhatkar, a brilliant Indian electronics technologist, was challenged by the government to develop an indigenous supercomputer, as a result of which 'Angry India' (as the *Washington Post* referred to our nation then) developed its own PARAM series of supercomputers.

Norman E. Borlaug had established the dwarf variety of high-yielding seeds in the 1950s. By the 1960s, Indian food production was in a mess and American agricultural economists had predicted starvation in India. In 1965, American President Lyndon Johnson had suspended the supply of foodgrains under PL-480 (the US Public Law 480, which approved the supply of wheat to India). The three S's—C. Subramaniam, M.S. Swaminathan and B. Sivaraman—acted on the charging-up question: Can India feed itself? India promptly responded by launching its historic Green Revolution.

John F. Kennedy spurred America to achieve the seemingly impossible feat of sending the first man to the moon. In organizations, leaders must create a charged-up environment for employees to take chances and succeed.

The *second* way to fire up associative thinking is by emphasizing not only the novelty that innovation brings but also its human impact.

Innovation can focus on the have-nots without compromising on profits. Well-known contemporary examples include the Narayana Health hospital, formerly known as Narayana Hrudayalaya, and Arvind Eye Hospitals, which conduct surgical operations at a fraction of the conventional cost and time. India is reported to have built a $3 billion forex-earning medical tourism industry.

In the late 2000s, the National Innovation Council (NInC) promoted inclusive innovation through State Innovation Councils, under which entrepreneurs and workers were being trained to innovate in areas such as fruit processing and leather through their 'innovation clusters'. An energy-saving furnace for brass-making has already been developed through the Moradabad cluster. Smokeless *chulha*s (traditional cooking device), distributed solar energy and low-cost cooling for food are all innovations that are waiting to be scaled up.

The *third* way relies on making teamwork a virtue. Holistic innovations usually span a whole value-chain and can be driven and executed only through a team representing various functions and domains. It is important that members for an innovation project be chosen for their ability to work in a team and not only for their individual brilliance. In fact, overtly competitive managers are less valued in associative thinking projects because they disturb the team dynamic.

The *fourth* way is to deliberately cut through the adipose tissue of organizational boundaries. Increasing socialization can be a very potent tool to create a culture of innovation. Such socialization makes the organization permeable and facilitates an exchange of ideas.

The *fifth* way is to wilfully unleash the creative forces within the organization. An executive's reputation and reward are based on mistake-free operations and not on the notion of the executive learning from failures. This severely restricts innovation. The process of making innovation an integral part of an organization's culture requires a willingness to accept creative failures.

Companies should celebrate success but should also celebrate authentic failures. Innovation can thrive only in a climate that encourages risk-taking. Institutions need storytellers who can repeatedly narrate inspiring examples. Innovation in human society is natural, just as nectar in food is natural. It needs to seep in everywhere, it needs to be warm to the touch and appealing to the senses.

Encouraging fresh thinking: Engineer Syamal Gupta

An idea is worth something only if you can sell it. Young executives feel that despite the bombast at company meetings, their leaders do not demonstrate an emotional entanglement with innovation to drive it. They state how their great suggestion did not get implemented, or how due promotion or recognition did not follow in some cases. They seem to expect their leader to behave like a hungry lion who should leap and grab at innovation with alacrity. The leadership view, on the other hand, is that young people must show innovation stamina. They must not just throw up ideas; they must develop the idea in detail, subject their proposal to challenge and review and, above all, must demonstrate a personal commitment to persist with the idea to the end.

As a result, most companies have a plethora of innovative ideas in the pipeline. An absence of innovation stamina relegates these ideas to PowerPoint presentations that are permanently awaiting execution, in turn, creating a negative spiral in the organization!

There is much emphasis on the culture of innovation, on the organizational atmosphere and attitudes and on the kind of visible leadership attributes that suggest an emotional entanglement with innovation. Stories from within an organization often say a lot about its culture because they demonstrate the right-brained facet compared to the left-brained survey data.

Here is one such incident from fifty years ago, scooped out of the archives of an Indian company, which has existed for over a century and continues to flourish even today.

In 1960, India, which could then barely feed its population competently, was not seen as being technologically savvy. Innovation was rarely discussed among the business community or economists. The top leaders of a large engineering company were in despair because they found that the equipment at a new and expensive plant exhibited 'several shortcomings—there was double handling of materials, the operation of the plant was slow and time-consuming, and the plant could not deliver its rated output', according to the archived documents. The production and engineering department were at loggerheads. The top management accorded high priority to solving the problem and the chief engineer was tasked to quickly resolve this major crisis. Such a situation arises these days as well, and typically, the task-force leader would nominate a crack team to solve the problem.

In this case, the records show that 'one person who interested himself was a young assistant engineer in the project department'.

Syamal Gupta was that young engineer. What does it mean to say he interested himself? Did he not wait to be nominated to the team? How could he be successful within such a traditional and hierarchical organization? He, however, did not seem to be bothered about having a mandate, reporting structure or resources. He moonlighted in his spare time.

Gupta, as subsequently recorded by chief engineer K.P. Mahalingam, 'gave considerable thought to the matter and evolved a proposal'. Almost all on his own! The records show that the seniors in the operating and engineering departments were taken aback and were initially dismissive. Thanks to the support of the chief engineer, as the archives reveal, 'The various aspects of the proposal were thoroughly studied with all concerned, and it was finally decided to adopt it. The financial savings were computed and found to be significant.' The internationally reputed German company Demag was the equipment supplier. The German engineers reviewed the innovative solution at the highest level and endorsed the proposal, thereby undertaking to manufacture the required new parts in their country.

A year later, a memo was sent by the chief engineer to the company's top leadership recommending Gupta for a modest reward, the nature of which they decided to 'leave to the senior management'. Gupta was noted as an innovator, but not decorated with a new office, designation or big increase in salary.

Gupta carried on as an assistant engineer. Two years later, he sought study-leave without pay to work at a German rolling-mill company and to acquire the Diploma of the Imperial College (London) at his own expense. Upon the completion of his diploma, he was commended once again by the chief engineer who, in a letter wrote that he was, 'the first employee of our company to acquire such a high qualification. With his attainments, a lucrative position in the UK or Germany was within his reach, but the fact that he elected to return to his company in India to re-assume his modest position is ample testimony of his loyalty and sincerity'. The chief engineer ended his letter with the suggestion that Gupta be considered for an increase in salary, retaining his role of assistant engineer. The suggestion was accepted.

Mahalingam rose to become director of technical services and Gupta went on to be director, Tata Sons. Gupta is now eighty-four, while Mahalingam is ninety-six, and the company they worked for was Tata Steel. This may be impossible to imagine in today's world, but it is not. Any leader can emulate these stalwarts within his or her unit or department. This is not a story about the individuals or the organization.

There are lessons.

First, that a company needs unstoppable youngsters as innovators; second, that these innovators must not merely be prolific in the generation of ideas, but must also have innovation stamina, or the follow-through skills of advocacy and persistence; third, that innovators should not seek instant gratification as a reward; and fourth and last, that the leaders of a company must be so emotionally entangled with innovation and the young people's aspirations that they do the job for which they are really paid— which is to groom younger people. This is all good old common sense, but it is conspicuous by its absence in far too many cases.

Of the four promoters that develop a concept into an idea— serendipity, mindset, associative thinking and articulation—the final one requires innovators to draw and sketch, describe verbally and write, whatever it is that is swimming around in their head. This involves rigour and discipline in developing thought leadership.

Articulation

Human connectivity with thought leadership is easy, a fact well-chronicled by Israeli historian Yuval Noah Harari in his writings. Humans are the only species—among the 8.7 million created—

that think, reflect, act, discover and learn in a cycle of incessant learning. To do these five things is distinctively human and natural and leads to individual learning, or what academics call 'implicit learning'. Now, this implicit learning gets converted into explicit learning through articulation. Some academics also use the expression 'thought leadership'.

I reflected on the origin of thought leadership. It was used for the first time in 1887 to describe Henry Ward Beecher, an American clergyman, social thinker and opponent of slavery. I also encountered this expression in the marketing literature of the 1980s when I was a marketer at Unilever. However, it was the obituary notice of the premature death of Joel Allen Kurtzman in April 2016, which brought back this expression into my consciousness. Kurtzman edited the *Wall Street Journal*, the *Harvard Business Review* and business magazine *strategy+business*. It was at *s+b* that he became credited with coining the expression 'thought leader' and of giving it literary visibility.

Institutionalizing thought leadership

The ancestor of innovation is thought. Without thought, there could be no innovation. In management, while innovation attracts profound commentary, analysis, awe and cynicism all at once, thought attracts less commentary.

You can imagine what a business veteran of fifty years like me thought when I got invited to a day-long workshop on the subject of thought leadership at an eponymous management college. My 'company executive' image about management faculty was shattered by an engaged and lively faculty, jousting and jostling to extract the best out of our discussions on a hot May day. It was well worth my time.

This workshop was hosted by Mumbai's prestigious SP Jain Institute of Management and Research (SPJIMR). Thought leadership at SPJIMR is the brainchild of the dynamic dean, Ranjan Banerjee. In the workshop, he was ably assisted by the equally impressive Snehal Shah—an ardent yoga student with a doctorate in organizational behaviour from Carnegie Mellon University.

Collectively, the faculty engaged on a journey—what is thought leadership, why is it important, what are the processes involved, what does a thought leader actually do? At the end of that day, it was not that the fifty assembled people had stumbled upon a revolutionary new idea; they had only emotionally engaged on a journey of self-discovery about themselves and their roles as management teachers—a hugely worthwhile outcome with, hopefully, more to come in the future.

You convert implicit into explicit learning through articulation. This is what teachers and managers do all the time, though they both do it differently, their efforts bring in their own value. Articulation can be through speaking, writing, demonstrating or through the ubiquitous PowerPoint-presentation technique. Teaching through articulation is a distinctively human event. Management teachers are in the business of converting implicit business knowledge into explicit, teachable learning.

Teachers who don't think that thought leadership is their core business are in danger of being considered lazy or indolent. Thought is distinctively human. Even intelligent animals like dogs don't conduct day-long seminars on thought leadership—only humans do such things in companies, universities and institutions!

I have observed the faculty being hard-pressed for time at management institutes, debating about what they should aim for—thought leadership or teaching. Institutions need to emphasize that teaching and thought leadership are not a zero-sum game. Both are essential—just like you have two eyes, and even though one may be stronger or weaker, the optician still has to fit you with glasses that allow the maximum usage of both eyes, individually and together.

The same is true of operating managers; that is why GE runs its management campus Crotonville, Unilever has Four Acres and Gulita, and Tata has the Tata Management Training Centre. I wish more power to thought-leadership initiatives. I hope many more management institutes as well as operating companies join this fray of thought-leadership because, without it, no institute can be innovative.

In a nutshell

In this chapter, I have likened the transformation of an embryo into a baby with how a concept in the brain transforms into an idea. While there are biological and natural phenomena that occur, as people interested in promoting creativity, we need to be aware of the factors of serendipity, the role of mindset, the importance of associative thinking and learning the discipline of thought leadership to stimulate the brain's successful conversion of a concept into an idea.

Reading about what happens to the embryo inside the womb triggers the imagination about what might be happening inside the brain to convert an embryo into an idea. I have been provoked to think about the parallels between a concept in the brain and an embryo in the womb by reading about this early stage of life.[8]

III

Infancy: A Prototype That Expresses the Idea

From birth till the age of about three, the baby becomes an infant. During this phase, the infant displays exceptional levels of curiosity—experimenting and probing all the time. Sometimes, the infant faces setbacks, but it takes risks by touching new things and observing its surroundings with wonderment. The infant is on a continuous learning path.

In the same way, an idea gets built up as a mock prototype, which has to be constantly modified and improved. Sometimes-harsh feedback about this prototype requires the innovator to keep an open mind and continue improving it. As a consequence, the prototype changes its physical manifestation, and the near-final prototype may bear little resemblance to the original version.

During this infancy stage of innovation, there are some promoters and accelerators of the process:

- Open-mindedness
- Overcoming setbacks
- Risk-taking
- Continuous learning

Open-mindedness

An open mind is the obverse of the coin of curiosity. Open-minded people are consequentially curious, and vice versa. Open-mindedness and creativity involve distinct habits of the mind and patterns of behaviour that must be cultivated.[1]

The important point to note is that creativity is not as genetic as may be popularly believed. It is a lot more an effect of the environment. Creativity is much more about setting expectations and getting people to behave in a particular way. If you explicitly or subtly 'classify' people, then they start to behave in line with that classification. For example, institutions tend to subtly classify people for administrative convenience. The creative class—advertising, marketing and research departments—are expected to work with imagination and originality. The other class—logistics, operations and finance—is expected to work with diligence and accuracy. In the due course of time, people start to behave in accordance with their classifications.

There is considerable anecdotal and scientific evidence that creativity is not in the genes:

i.   When Albert Einstein died, researchers opened up his brain and, contrary to the expectation of a humungous brain or special brain bumps that might explain his genius, researchers found that Einstein's brain was actually smaller than that of most ordinary people.

ii.   A research study was conducted by the department of psychology at the University of Bronx to investigate the creative abilities of identical and fraternal twins in 1973.[2] The aim was to establish the connection between genes and creativity. The researchers studied all the twins born in Connecticut since 1897. Although the genetic composition of similar and dissimilar twins is different, the researchers found no difference in the rates of creativity between the two types of twins.

Curiosity is the 'engine of intellectual achievement and it is what drives us to keep learning and pushing forward', according to science writer Annie Murphy Paul. No sooner is the child born that it displays intense curiosity about everything. In the same way, every articulated idea gets nurtured and modified by curiosity. Curiosity is the essential hormone for creativity and innovation.

We are not sure why we use the expression, 'Curiosity killed the cat'. In 1588 British playwright Ben Jonson used the expression, 'care'll kill a cat'. In his play *Much Ado about Nothing*, William Shakespeare wrote, 'What, courage man! what though care killed a cat, thou hast mettle enough in thee to kill care.' By 1873, the expression morphed into 'Curiosity killed the cat.'

If the truth be told, curiosity is the preserve of all forms of genius. It is curiosity that accounts for the progress of mankind. We should recall what was written by Einstein: 'I have no special talents. I am only passionately curious.'

'Curiosity is one of those personality traits that gets short scientific shrift,' according to an article in *Wired* magazine.[3] Raw intelligence has been researched to death, but our curiosity about the world remains a mystery. In a seminal paper published in 1994, Prof. George Loewenstein argued that curiosity comes alive when 'we feel a gap between what we know and what we want to know'.[4]

What a marvellous insight—so simple, yet so powerful! It is a different way of saying that everything humankind knows so far (undoubtedly an impressive amount) appears like a set of little dots on an enormous, white surface of ignorance. The statement makes knowledge humble.

So if you know absolutely nothing about a subject, you have no curiosity. If you feel, however, that you know everything about a subject, then your curiosity about knowing more becomes zero. In between knowing nothing and everything lies the point where you know a little, but do not believe that you know too much. If knowledge is on the x-axis and curiosity on the y-axis, then curiosity peaks in an inverted-u shape. This was scientifically established by the laboratory experiments of Prof. Colin F. Camerer at the California Institute of Technology.

Scholars suggest practical ways to stimulate curiosity. First, teachers should spend time asking the right question, one that opens up an information gap in the student's mind. An alumnus of the London School of Economics, B.K. Nehru, apologetically told his former teacher Harold Lasky that, going by his own civil service experience, the latter's ideas did not work in India. Lasky responded, 'But I gave you no answers. I was merely teaching you how to think.' Second, be aware that you know only a little bit because then you want to know more and your curiosity pump gets primed. Third, communicate with others, because language and communication are the tools of natural learners.

Creativity at the firm level

I like the charming story that follows, to support the view that open-mindedness and curiosity are cultivated virtues.

In 2003, an attempt was made to find 'the most innovative company in America'.[5] That was quite a challenge—you could count the dollars spent on research and development or the number of patents obtained, but as an alternative, you could look for a company that had a long and consistent record of innovation. The investigators and researchers looked for a company where innovation was resilient and did not depend on the genius of one individual. Surely that is a good definition of an innovative company! They stumbled upon W.L. Gore & Associates. Who are they?

They are a company that is 5000 kilometres away from Silicon Valley and even further away from Wall Street. It is privately owned and enjoys a double-digit revenue growth every year. The company employs over 6000 people and is stated to bring in about $3 billion in revenues. Its cutting-edge fabrics are worn by astronauts and soldiers, its heart patches and synthetic blood vessels have been implanted in about 8 million patients.

And what have they done?

The founder of W.L. Gore & Associates quit a long career at DuPont with the specific intent of founding a creative company in 1958. In the next fifty years, W.L. Gore generated $3 billion in revenue from a diverse set of products. The founder organized the company so that everyone could be creative. For example, David Myers, an engineer working on cardiac implants, became curious about how to improve the performance of the cables on his mountain bike. His openness to experience resulted in his developing a suitable polymer, branded GORE-TEX. This polymer was used by the company to coat guitar strings for better tonal quality!

Here is another way to learn and develop open-mindedness and curiosity.

Lessons from Jesuits

I studied at St Xavier's Collegiate School in Kolkata, a Belgian Jesuit institution. The dedication as well as the curiosity, spirit of inquiry and erudition of the Jesuits have always fascinated me. They seemed deeply immersed in whatever it is they were studying, whether it was physics, archaeology or even Sanskrit. Over the years, I have tried to understand this order with specific regard to creativity and innovation. Their origin and evolution bear lessons.

The Roman Catholic order of the Society of Jesus was founded by the Spaniard St Ignatius of Loyola, along with nine other companions in approximately 1540. One of those companions was St Francis Xavier, whose exposition in Old Goa is currently on. He arrived in Goa in 1541 and held the record for the highest number of Christian conversions during his time.

The early Jesuits realized that the hierarchical Catholic Church was in dire need of reform. The clergy then was poorly educated. The Jesuits were 'men on the move', ready to go anywhere on a mission under obedience to the Pope. Education was not the principal goal of the early Jesuits. Soon, the founders realized that intellectual competence was essential to bring about change. During his lifetime, Ignatius opened as many as thirty-three schools. The Jesuits insisted on a high level of academic preparation for those wishing to be ordained into the ministry. Knowledge became an enabling prerequisite. That is how, although the order had received papal approval in 1537 and despite their loyalty, Ignatius and his successors periodically upset the Pope and the church bureaucracy. I reckon this happens to all creative people.

The central tool used by the Jesuits to change people's hearts and minds was the 'Ignatius retreat', or spiritual exercises. This required a four-week period of silence, directed meditations and conversations with a spiritual director. The retreat was aimed at ridding a person of predispositions and biases, enabling the person to make free choices. It was through this technique that Jesuit education burst forth on to the world in the sixteenth century.

The techniques of silence and meditation are not unique to the Jesuits. They are well known in yoga as well. Thinker and author Pico Iyer spoke eloquently on the subject:[6] '[F]reedom from information, the chance to sit still, that feels like the ultimate prize.' The lesson is that you need a method to develop an open mind.

Armed with knowledge and an open mind, the Jesuits became very focused and disciplined about their subjects of study. They learned how to specialize in a subject, yet integrate their knowledge into their theological base. Deep specialization rested on a philosophical platform of education and innovation.

The results were an effulgence of innovative thinking and learning. The Jesuits made such significant contributions to the understanding of earthquakes that seismology was even described as 'a Jesuit science'. During the seventeenth century Jesuits made important contributions to experimental physics. In the glittering Chinese Ming courts, the Jesuits were regarded as 'impressive for their knowledge of astronomy, making of calendars, mathematics, hydraulics and geography'.

Four lessons for modern innovation can be drawn: first, recognize that knowledge is essential; second, adopt a formal method for the development of an open mind; third, learn the art of making free choices; and fourth, specialize deeply on a philosophical platform.

We somehow assume that we require no training in our 'natural functions' like breathing, eating, concentrating and innovating. Yoga teaches us that we do need such training. Maybe it is time to introduce vipassana into innovation training in companies.

Moments that boost our creativity abound in every sphere and always arise out of behaviour. There are four types of creative behaviour: first, an insatiable curiosity leading the person from one unexpected discovery to another; second, taking a break, thus avoiding being overtly fixated on the problem; third, a determined journey through the five stages of creativity, described by Hungarian psychologist Mihaly Csikszentmihalyi as preparation, incubation, insight, evaluation and elaboration; and fourth, resilience to cope with the periodic feeling of not getting anywhere. These four apply to science and technology as much as to writing and the performing arts as demonstrated by the following story of an accidental writer, fired with curiosity.

The trail of the last king of Burma

Sudha Shah had never written a book. She is a first-time author, but wrote an outstanding book.[7] I interviewed Shah to find out how she conceived the idea of a book on such an obscure subject.

Insatiable curiosity: It was Amitav Ghosh's very evocative book, *The Glass Palace*, which first piqued Shah's interest in the last king of Burma and his family and got her started on an extraordinary journey. Not much had been written about King Thibaw and his descendants after his exile. Shah's book is about the king, his wives, his daughters and his grandchildren. It has not been written as a work of fiction but as a family biography. Part one of the book depicts the life of the king and queen during their reign. She felt that this background was essential to put into context the royal family's life during and after the exile. Parts two and three are really the heart of her research and of the book— they detail the family's life during and after the exile. Although she has written the book as a human-interest story, it is set in its historical, political, social and cultural context, for without this, it would not have had much meaning.

Journey through stages: She had absolutely no intention of writing a book, but simply wanted to learn more about the king's four daughters, who grew up in the culturally alien environment of Ratnagiri, with virtually no education and with almost no interaction with the outside world.

Resilience: Shah was deeply frustrated by the absence of records and artefacts, which she could obtain only after a lot of delving and persistence. She became fascinated with not just the princesses but also with the other family members and aspects of the story. It just snowballed; she certainly had no idea that it would become such an obsession that she would spend over seven years on the subject with enough breaks from the intense trail!

Avoiding over-fixation: She looked for a publisher only when her manuscript was almost ready, and she was, therefore, not pressured by a deadline. Shah says that she felt like a detective. Tracking down a variety of sources and accessing new information, however insignificant, was always very exciting! It was the details that helped bring the protagonists alive for her. She took her time to research with some essential breaks, so as to craft a strong narrative.

When she began her book, her objective was to uncover and describe the lives of the four princesses. However, as her research progressed, the focus of inquiry of the book shifted to provide an insight into, first, how an all-powerful and very wealthy family coped with forced isolation and separation from all that they had once known and cherished and, second, how the royal family coped once the exile ended and they were abruptly released into a world they knew almost nothing about.

Through her open-mindedness and curiosity, Shah produced a very creative book. The message is simple: like the infant, stay open to new experiences, be permanently passionate and curious and always allow yourself to be led through the journey indicated by your curiosity.

Overcoming setbacks

My curiosity was roused by a Chinese-American visitor who told me, 'Every Indian has a rich environment for creativity—his or her life faces all the C's required to be creative—chaos, challenges, curiosity, communications and channelization. Such life challenges provoke curiosity and creativity.'

'Yes,' writes Modupe Akinola, an assistant professor at the Sanford C. Bernstein & Co. Center for Leadership and Ethics,

'Fostering creativity is vital to the modern economy, but to reach your personal best, sometimes you have to go through your worst.'[8]

Anecdotal evidence links creativity to suffering. Think of the ancient Greek musician and poet Orpheus, or the nineteenth-century painter Van Gogh, or modern-day singer Amy Winehouse. Philosopher Friedrich Wilhelm Nietzsche was a tortured soul who believed that suffering spurred his creativity; Blaise Pascal, philosopher and mathematician, suffered ill health. Poet Lord Byron wrote, 'An addiction to poetry is very generally the result of an uneasy mind in an uneasy body.'

Does pain promote creativity?

NephroPlus

Modupe Akinola's experiments suggest that those who receive a sharp critique of their work may become emotionally vulnerable and go into depression, thereby developing low levels of a hormone called dehydroepiandrosterone sulfate, or DHEAS. This hormone, however, in turn, promotes high creativity!

I found Akinola's words rather intriguing. While I mulled over her paper and its averment with curiosity, I met Kamal Shah in Hyderabad. He is a strapping, forty-something chemical engineer who, along with Vikram Vuppala and Sandeep Gudibanda, runs a successful start-up called NephroPlus, India's largest chain of dialysis centres. The company is about seven years old, makes a profit (as normal businesses are supposed to) and caters to about 7000 dialysis customers every year, of whom a fourth depend on public–private partnerships to fund their treatment.

Almost like a dreamy visionary, Kamal Shah asserts, 'Our aim is that our dialysis guests (not patients!) should radiate positive energy and be strong in the self-belief that they can lead a full and normal life. At NephroPlus, we combine technology, ethics and guest-centricity.' Kamal Shah is a self-assured and passionate entrepreneur who has already attracted three rounds of funding from investors. So is this just another Gujarati guy who has once again demonstrated the fabled entrepreneurial DNA? No, he leapt forward from a potentially debilitating health setback, almost as a walking proof of Akinola's hypothesis.

Cut to 14 July 1997. A gold medallist chemical engineer prepared furiously to go to the US for his master's degree in science. To get the visa, Shah had to get a vaccination, which threw up an unknown and rare setback—his kidneys were not functioning. How could that be? He was a superactive, healthy youngster. He and his supportive family assumed that there had been an aberration, which would soon pass. Over the next few years, however, Shah went through hopeless treatments, severe pain, emotional trauma, a failed kidney transplant from his doting mother and an infection arising from exposure to the 2004 Chennai tsunami. Driven by curiosity, Shah researched and educated himself on the subject, only to face the harsh reality: he would have to depend on dialysis for the rest of his life.

He did not measure his DHEAS hormone, but it could well have been low. In a leap of faith and curiosity, he started to blog about his experiences with other dialysis patients. Soon, he received an unsolicited message from entrepreneur Vuppala, who roped in Gudibanda. The three youngsters brainstormed on whether they could provide a happy solution rather than watch the world's largest dialysis-stricken nation fail itself psychologically. Yes, India is known as the diabetes capital of the world. It is also the nation that needs dialysis the most in the world! The greatest asset of this team was that none of them knew anything about healthcare or dialysis. They were just driven by unbridled curiosity arising out of their knowledge gap.

From the patient or customer viewpoint, traditional dialysis is perceived as a huge impediment because it limits one's movements, activities and makes one prone to infection. Online research fired the neurons in the trio's brains and implanted a concept into their minds: Why not build a network of dialysis centres that puts the guest at the centre of the universe, where they are enabled to live a full, productive and long life? That would require protocol to be thoroughly followed and processes to be in place as well as a cadre of trained dialysis technicians to be present. They went right ahead with their passion for the idea and soon got some of India's best nephrologists on board and formalized the protocols.

There is an acute shortage of well-trained, qualified and ethical dialysis nurses and technicians. That did not deter them. On the contrary, they said to themselves: NephroPlus would produce such a cadre. Enpidia, NephroPlus's training division was set up to recruit nurses and other undergraduates and graduates to train them to handle dialysis guests. Over 350 Enpidia technicians support the NephroPlus dialysis centres as a living testimony to skill building in India.

It is boring for the guest to sit around waiting for the hours required to complete the dialysis; so NephroPlus centres provide a television, Wi-Fi and Internet access to keep the guests occupied. A pickup and drop facility is also arranged so that guests can come independently to the centre without depending on a family member.

Listening to Kamal Shah's story does convince you that while nobody consciously seeks setbacks, if one does happen, you can take a creative leap forward.

Conseillers en Gestion et Informatique Corporation

There is another illustrative story about Canadian Serge Godin, whose family sawmill business was destroyed by fire. Despite that setback, Godin grew up to set up a multi-billion-dollar computer business—Conseillers en Gestion et Informatique Corporation (CGI).

In 1966, Godin was just seventeen years old and living in rural Quebec, Canada. The family had a sawmill business. The sawmill was uninsured and, in a devastating fire, was completely burned down. All of a sudden, the family required everybody to start working and bringing home some earnings. Serge worked in a supermarket and then at a drycleaner's store. The setback got his creative juices flowing: he was determined to do something different and never face such a ruinous situation again.

By the time he was in his early twenties, he had C$5000. He started CGI in a brand-new risky domain. A garage seems to be an essential starting point for tech companies, and sure enough, Godin too began his business in a garage.

Over the years, through trial and persistence, CGI has now become an enterprise worth C$10 billion. With a personal wealth of C$1.5 billion, Godin reminisces that he has built a people-centric company, whose profits provide the inspiration for his charitable institution, Fondation Jeunesse-Vie, which means 'youth foundation'.

Setbacks do teach you lessons, provided you are prepared to learn—like infants do.

Jamsetji N. Tata

Business managers are essentially men and women of ideas and curiosity; business is not far from the world of ideas unless you assume that to be so. Commerce and business are mere manifestations of the dreams and passions of people and organizations. Having served Tata for many years, I was intrigued by what made the founder, Jamsetji N. Tata, pursue all that he did pursue. Was it raw passion or inspiration from ideas, or both?

Listening to a speech by influential British author and philosopher Thomas Carlyle—'the nation that has steel will have gold'—sent Jamsetji on a voyage of curiosity and the subsequent establishment of a steel business. Long sea voyages and relaxed conversations promoted curiosity. A chance meeting with Swami Vivekananda on a ship from Yokohama to San Francisco in 1893 fired up Jamsetji's mind with an idea that later manifested itself as the Indian Institute of Science.

The legendary J.R.D. Tata, in his foreword to Frank Harris's biography on Jamsetji N. Tata in June 1957, wrote, '[M]en of business are not often at home in the world of ideas; it was Jamsetji's distinction that he lived in both worlds, the world of ideas and the world of action. It was because Jamsetji Tata lived in the world of ideas and had imagination that he played the role of pioneer in India . . .'

Jamsetji's zest for life and humanity kept him curious.

Risk-taking

It is fascinating to watch how infants take innocent risks and constantly learn from the outcome of the risk taken. Innovators do the same thing. They don't know exactly what lies ahead, but they take a measured risk to find out. The field of archaeology provides some examples.

Mohenjo-Daro

Archaeologist Sarah Parcak says that when they are digging, archaeologists dig for people, not things.[9] 'Never are we as present as when we are in the midst of the great past,' she says. Sarah is not just any ordinary archaeologist, she is a space archaeologist. This means that with the data that becomes available from satellite images of very fine resolution, she processes these images using algorithms. She can see things that are buried under the ground and that enables her to go and dig in a productive manner. 'Good science, imagination and a leap of faith are the trifecta that we use to raise the dead,' she says evocatively. And, in its own way, that is what the infant does, as indeed the innovator.

My great-grandfather's generation had no knowledge of the existence of Pataliputra and Mohenjo-Daro. Both archaeological sites were discovered only in the early twentieth century. Two recent books tell the story of serendipity.[10]

By the end of the nineteenth century, Sir William Jones had decoded an unread Sanskrit text *Mudra Rakshasa* and established that the Greek reference to the Erranoboas river was actually to the modern Sone river. This accident resulted in our learning the story of the Maurya Empire. In 1901, Englishman John Hubert Marshall, who was trained in the Greek classics and who had never travelled beyond the Mediterranean, was appointed to a senior post in the Archaeological Survey of India. Over the next twenty-two years, he discovered the connection between the Harappa seals and Mohenjo-Daro, as also the ruins of Pataliputra, which incidentally was made possible by a grant from Sir Ratan Tata, Jamsetji Tata's son.

ALZUMAb from Biocon

Though Ms Kiran Mazumdar-Shaw is known for assiduously building up the public company Biocon, lesser known is her vision to discover something truly different. Five years ago, Biocon announced a world-class drug, branded as ALZUMAb for the treatment of psoriasis—a chronic skin condition that causes the skin to flake off, leading to redness and irritation. I did not know that its story contained a risk and a fortunate coincidence. I wanted to find out more. Cuba and pharmaceutical research were not connected in my mind.

Che Guevara, who was a political revolutionary and physician, wrote his eponymous paper on revolutionary medicine in 1960. Following the revolution and American embargo, the Cuban government adopted universal healthcare as a state priority and committed big funds for research, though their success in exploiting said research has been less than clear.

In 2001, the science-oriented and commercially savvy Kiran Mazumdar-Shaw visited the Institute of Hematology and Immunology in Cuba. She spotted a molecule which showed promise in treating autoimmune diseases. Normal immune cells recognize self and non-self; they are harmonious with the self and attack the non-self. However, an aberration occurs when the same immune cells attack their own kin. This condition of attacking the self leads to autoimmune diseases like diabetes, rheumatoid arthritis and psoriasis.

This magic molecule, christened T1h, showed promising results with early-stage discovery. The Cubans licensed out the molecule to Biocon for further development. For the next ten years, Biocon pursued a drug development agenda, based on the then well-known scientific principle of inhibiting the Th1-mediated immunity called the 'Th1 pathway'. But they encountered some inexplicable results during their work. The molecule appeared to work through mechanisms other than the Th1 pathway too. For Biocon, the penny dropped when they noticed a 2009 scientific paper on a Th17 pathway, which postulated a new pathway, involving different cytokines than those that were part of the Th1 pathway.

Suddenly, Biocon's scientific observations of the last ten years seemed to make sense, and the cytokines involved seemed to explain the action of ALZUMAb—the inexplicable had an explanation and the intriguing facts became rational!

Biocon had to move quickly. Since the company had done the development work on psoriasis, Biocon used its extensive trial data to get approval for psoriasis within a short time.

On 10 August 2013, Mazumdar-Shaw dramatically announced to an excited audience that the company had a first-in-class drug that could well be a best-in-class drug. It was a breakthrough, because ALZUMAb would be a 'biologic': it would not be made by chemicals, but would be a natural molecule made up of protein and sugar. It would have lesser side-effects and would be able to prevent the disease from recurring for a prolonged period. Four weeks later, on 9 September, Orca Pharmaceuticals, a new company based in the US, was announced to develop biologics for psoriasis. The race is on.

Quite a bagful of risks and coincidences, isn't it? But let us not miss the four lessons. First, that Mazumdar-Shaw kept an open mind to strange, new ideas by fishing for them in Cuba; second, that she was focused and disciplined by exploring T1h for ten long years; third, that when a new insight fortuitously landed in its lap, Biocon moved fast to change; and fourth and last, the company paid attention to intellectual property rights protection as well as product branding. They almost seemed ready for luck!

But then again, as the saying goes, luck favours those who prepare for luck by continuously learning!

Continuous learning

Learning and innovation are companions, and the stories of their companionship are interesting and inspirational. We fantasize that there exists that elusive Eureka moment because of the stories we have heard of Isaac Newton (gravity) and Alexander Fleming (penicillin). We hope that Eureka moments like these happen to us as well. It is that one moment when you sense that all the pieces have come together. It is usually in the incubation phase that you get that Eureka moment.

On the other hand, from stories such as Biocon's ALZUMAb and Hindustan Lever's Fair & Lovely, it also seems that there is some mysterious process called 'mental preparedness'.

So how can you develop such mental preparedness? French mathematician Henri Poincaré had enunciated four stages in this regard: preparation, incubation, illumination and verification.

Almost 100 years after Poincaré, psychologist Mihaly Csikszentmihalyi researched this aspect by studying the thought processes of people who are acknowledged to be creative, rather than the usual research of investigating the inner workings of their minds. He postulated that there are five mental stages: preparation, incubation, insight, evaluation and elaboration.

Psychology research throws up some interesting data.[11] Does it matter a lot whether you concentrate or face interruptions in your process of cogitation? In one experiment at the University of Sydney, three groups of subjects were asked to generate ideas on the alternate uses for a piece of paper: the first group was allowed to work on the problem, uninterrupted, for four minutes; the second group had one interruption at the end of two minutes on a related subject; and the third group worked for four minutes with a half-way interruption on an unrelated subject. Guess which group produced the most ideas? The last group! It was the group that was not in a position to concentrate during the experiment.

So much for people who complain that their job leaves them no time to think! At the University of California, Santa Barbara, an experiment was designed to study if it was focused minds or wandering minds that produced more ideas. It was found that wandering minds generated more ideas.

In such a creative or wandering mind, the neurons are firing furiously but not in any orchestrated manner. Because the mind goes superfast through all the five Mihaly stages, it is concurrently evaluating and elaborating those ideas, and trying very hard to develop an insight. When that insight develops, the 'neuron orchestra' gets locked into a rhythm, and that is the birth of a human idea.

Tata Swach

TCS had recruited as a consultant one Prof. P.C. Kapur, who had retired from the Indian Institute of Technology Kanpur and who had done some pioneering work on waste materials, including on the adsorptive and antimicrobial properties of rice husk ash (RHA). His novel insight was that RHA had around 85 per cent activated silica and 10 per cent activated carbon. Thus, it could adsorb bacteria (due to the activated silica) and also remove odour (due to the activated carbon). RHA is mesoporous, which means that it has pores within pores. Hence, its adsorptive capacity is high. Since its main constituents are in an activated state, it is possible to embed other substances on to it.

However, the problem was that this process removed only 85 per cent of the bacteria, apart from odours and suspended particles. As constructed in the laboratory, the product was clunky and useful only for emergency relief in cyclone- or earthquake-hit areas.

So the natural questions arose? Could the same idea be developed as a consumer product for 'just before consumption' use? How could the remaining bacteria and other disease-bearing viruses be removed?

I had learnt long ago from my grandfather that water stored overnight in a silver vessel was 'good water'. He had no idea why. Much later, when I heard about silver with respect to water, my brain could recall the old lesson and my neurons started to fire. With a wee bit of reading and some help from Google, I learned that silver has long been known as a metal that killed bacteria upon reasonable physical contact. Surely, one could not make bacteria-free water by placing burnt RHA in a silver vessel! My mind was a bit restless because I did not know how to connect native wisdom to modern science.

I learnt that one TCS scientist had proposed nanosilver, but could not figure out a safe and cost-efficient way to use the material. I jousted with Murali Sastry, who was then with Tata Chemicals. Could he throw some light on how silver vessels purified water? Of course he could. He rattled off some layman's introduction and some technical gobbledygook. Sastry said, 'Why think of silver when you can consider nanosilver? Let me do some work and get back to you.'

This meant that we could use very small quantities of finely particulate, nanosized silver. But it was quite unclear how this fine silver could be mixed with burnt RHA; further, the filter bed thus produced had to hold together as water flowed through it. 'Nanotechnology is generally used in very high-end products like cosmetics, because there is a cost involved in creating additional surface area. The scientific challenge was to do so at a low cost,' said Sastry.

A core team was created comprising scientists and managers from TCS and Tata Chemicals. The challenge was posed to them in lofty terms: Could the team do for drinking water what Thomas Edison had done for lighting by developing the electric bulb? The neurons of several people fell into some lockstep progressively. It took three years of consistent development work by a hugely talented team of passionate scientists, persistent development folks, adaptive production managers and compulsive marketing evangelists to design and launch Tata Swach, a novel nanotechnology-based water purifier.

It is fallacious to think that genes favour only some people to be creative. Serendipity and innovativeness don't come easy to some people because their genes predispose them to such an advantage. It is well established that creativity is not in the genes, but in the mindset and one's behaviour.

Through failed attempts and learning, the infant also learns self-control without loss to its self-esteem, what is also the beginning of thought mastery.

In a nutshell

Just as babies grow into being infants between birth and the age of about five, ideas develop into prototypes soon after the idea has been articulated by the person who had a concept in his or her brain. Getting trained to nurture during this phase is greatly helpful. Open-mindedness and curiosity are virtues to be nurtured; setbacks will occur but recovery must follow; measured risks need to be taken; and above all, the virtuous cycle of learning and articulation must be in play.

If this stage of growth of an idea is crossed successfully, then a prototype is ready to be converted into a market model for the customer.

IV

Childhood: A Model Shaped
Out of This Prototype

The prototype stage in the innovation journey is the point at which an idea is presented in a physical and working form, just like a child growing up with a personality and expression of its own.

As stated before, many concepts fail to become ideas, and many ideas fail to become prototypes. Leonardo da Vinci imagined a heavier-than-air flying machine, and though he could sketch what he had imagined as early as 1519, he could not construct it in reality. It was only in 1903, almost 400 years later, that the Wright brothers of the US managed to demonstrate a flying machine in Kitty Hawk, North Carolina. Their prototype can be considered the forerunner of the innovation that we recognize today as the aeroplane.

The creation of a working and consumer-friendly model is what this fourth stage is about. Of course, the model so produced becomes the basis for continuous and dramatic improvements for decades thereafter. If one uses the metaphor of the human being, the period when an infant grows into adolescence, approximately from the age of five till fifteen, would correspond to this stage—from prototype to model.

This is a frustrating phase for the innovator, just like it is for the parent who is raising the child. At this point, the four nurturers of the childhood stage are:

- Learning from failure
- Adapting continuously
- Coping with opposition
- Developing the story

Surely, there will be criticism of prototypes and their performance failures, and the innovator needs resilience to cope with failure. The continuous adaptation of the prototype into a working model seems to go on and on; only patience and self-belief can help the innovator get through this phase. The innovator encounters many opponents or non-believers; they will comment discouragingly about the innovation—a bit like a kibitzer (the Yiddish word for a person who keeps commenting cynically or negatively about why something would 'not' work, rather than actually contributing to making it work)! The innovator must persist in developing the narrative or story of the innovation. He must behave like a salesman who deeply believes that if others are not buying his idea, it is his fault; he has to learn how to tell the 'story' of his innovation.

Learning from failure

Dare to try

When an organization wants to be more accepting of failure and experimental, what are the barriers? Everyone agrees that breakthrough attempts are essential for innovation and, hence, the risk of failure must be accepted by organizations.

What everyone sees in reality is that organizational rewards go only to those who deliver fault-free work. There is even academic research to evidence this fact. For example, the creative folks in advertising are often given the caveat: 'No risks, please.' In an international company, the local team will not risk advertising anything that will be disapproved by those involved in judging these advertisements at the headquarters. Only occasionally is the barrier transgressed—like with J&J's 'Nude Models Wanted' advertisement by Trikaya Grey in the 1970s, or the whacky Wheel advertising by Hindustan Lever in the mid-1980s, and the more recent Cadbury bitter-chocolate campaign.

Large organizations think that encouraging failures would have implications on its reputation and thus want to play it safe. An organizational culture of taking small bets or making fatal errors at a later stage of that work can prevent the development of a weak culture. Most people are very unclear about how to accomplish this magical state.

Here is a way to think about the issue and classify error types.

Three types of errors occur in organizations: first, the error arising due to sabotage or intentional concealment. Leadership has to demonstrate zero tolerance for this type. Second, an error occurs due to carelessness or the bending of rules. Organizations develop some tolerance for these, but do not explicitly approve of their occurrence. The third type is a creative error, which occurs due to changing market circumstances, calculated risks, experimentation or even bad luck. These are 'good errors' and can be encouraged if someone defines what a creative error is. Only then can creativity be encouraged.[1]

Gerhard Bihl was the human resources director of BMW's Regensburg factory around 1990. His challenge was to convert the ideas from employees' craniums into practice. He began an activity called 'Flop of the Month', or, more elegantly put, 'Creative Error of the Month'. In contrast to the conventional 'Employee of the Month' scheme, which eulogizes the error-free, highly efficient and ideal employee, this activity focuses on the tragic hero of everyday business, whose experiences harbour unexpected learning potential. Bihl piloted the scheme in the Regensburg factory for three-and-a-half years, during which time twelve awards were given away.

In due course, the scheme petered away and the issue of encouraging creative errors remained a challenge.

In the US, Ireland and Sweden, the sharing of mistakes and lessons from misadventures is celebrated through what is called the 'Golden Egg' awards. As one member of an Ann Arbor association of corporate presidents puts it, 'I want to hear it from the member who got egg on his face trying out his idea.' The presentation of the award for the best mistake of the month became a standard part of their meeting, and the trophy itself added an important new dimension. It gave the company president the chance to be a model for treating mistakes as opportunities to learn how to do it better, rather than treating it as a situation requiring blame. It legitimized the importance of learning from both our failures and successes.

Tata has experimented with these concepts through a 'Dare to Try' initiative. A satisfying outcome has been the openness of managers to come forward with stories that were not successful. When this category was initiated in 2007, the Tata companies had to be cajoled to participate and there were only twelve cases from six companies! In 2016 there were 250 cases from thirty-five companies. The transparency in discussing such cases has helped in building a learning culture in Tata.

Three types of lessons have been learnt from creative errors:

i. *Technology advancement*: These attempts have helped teams uncover blind spots on the road to technology progression and have helped them take that big step through a risky project.
ii. *Business models*: Not all innovative products could be commercialized through the business models prevalent in

the company. Often, products with cutting-edge technology need to be supported with novel business-models.

iii. *End consumers*: In case of breakthrough innovations, the perception of end consumers and their consumer experience while using the product are extremely critical.

If the acceptance of creative errors is encouraged, teams and individuals will surely develop a learning culture, which will be a stepping stone for successful innovations. Such a discriminating approach to various types of organizational errors can help encourage the right errors and foster that elusive spirit of risk-taking that all organizations strive for.

NASA and the Hubble failure

A very sound technical team launched the National Aeronautics and Space Administration's, or NASA's, Hubble Space Telescope in 1990, unfortunately, with a flawed mirror.[2] A trivial and avoidable error overshadowed the accomplishments of thousands of dedicated people, in the process squandering $1.7 billion of taxpayer's money. Here is what happened.

After a textbook launch, the team soon discovered the flawed mirror. A detailed investigation followed and the findings were startling. A huge error was discovered in adjusting the null corrector used to figure the mirror that caused the flaw. The device was at the contractor's plant. Hints of the flaw in the mirror had showed up in numerous tests. The review board wondered why such a smart technical team had not rigorously pursued these hints. It was found that the schedule and budget pressures had caused them to move relentlessly forward.

The question was why had the NASA scientists and engineers not addressed these inconsistencies. The board then made a disturbing discovery. The contractor had never forwarded these troubling results to NASA! The board finally concluded that a leadership failure had caused the flawed mirror in the $1.7 billion telescope. NASA's management of its contractors had been so hostile that they would not report technical problems if they could rationalize them. They were simply tired of the hostility. Does this episode not remind the reader of experiences in his or her company?

Here is another example. Korean Air faced frequent air crashes between 1988 and 1998.[3] The average flight captain's social status was so high in Korean society that the junior officer would, at best, be oblique, even in those cases that required more direct communication. Thus, in most cases, this lack of communication and teamwork between the pilot and co-pilot—due to the former's power and high social standing—led to these plane crashes, as the co-pilots allowed the pilot to take all major decisions, even when these decisions were questionable.

Charles Pellerin explains a 4D team-building process including an online behavioural assessment to help understand each other and measure the key driver of team performance: the social context. Mature companies understand the impact of culture and team climate and have created mechanisms to measure and monitor these soft indicators which practically impact everything, especially the fostering of innovation. Companies need to inspire employees to put their most creative foot forward and come up with new ideas, concepts, processes, inventions or improvements.

Adapting continuously

Harnessing solar power

Attempts to use solar power have a long and meaningful history. Environmentalism and technological developments in using electricity from the sun make this subject the darling of venture capitalists. Breathtaking 'aha' manifestations in general are invariably consumer-led innovations. In the case of solar energy, they are adaptations of solar and electrical technologies to distinctive applications. Thus, consumer benefit becomes the central driver of innovation.

Here are four very different manifestations of the generic innovation referred to as electricity from solar power. But first, here is a quick introduction to some jargon: solar converts to DC (direct current) electricity. Our grid supplies AC (alternating current) electricity. Solar DC must be converted to AC through a device called an inverter unless the user's appliances are converted from AC to DC.

Laura Stachel, the creator of the 'solar suitcase' was shocked when she witnessed the poor obstetric care in a Nigerian hospital with unreliable electricity. She developed a portable, compact version of the hospital solar–electric system that could scale up to rural hospitals and clinics. Since 2009, Stachel has produced over 400 solar suitcases that have served rural hospitals in over twenty countries. Each system costs $1500 and takes an hour to install. Stachel, from the Blum Centre for Developing Economies at the University of California, Berkeley, was named one of the CNN Heroes of 2013. The solar suitcase is seen as an important innovation in the fight against maternal mortality in the world.

In the second example, the intrepid Prof. Ashok Jhunjhunwala of IIT Chennai observed that Indians suffer electricity blackouts. He questioned: Is it possible to avoid a blackout (the complete unavailability of power) and restrict the consumer's inconvenience to a brownout (the limited availability of power)? His question is simple, and the solution can be hugely impactful.

Blackouts occur when demand exceeds supply. So the innovators asked, instead of carrying out 100 per cent load-shedding, can, say, 7 per cent of power be allowed to continue to flow on the grid, albeit at a much lower voltage to distinguish it from normal supply? This limited power is distributed to homes in the DC form. Incidentally, DC appliances like LED lights, DC fans and DC electronics are also more energy-efficient, stretching far whatever limited power is available. The team figured that the grid could terminate at each home on a unit, which would now provide two power lines to the home. One would be the currently used AC supply, which could be used to whatever extent a customer wants, but would trip during load-shedding; the other would be a DC line with limited power, but available during normal supply as well as during load-shedding. Customers would use the DC line with DC appliances like lights, fans and electronics.

While the utilities will provide limited DC power, a customer could also directly connect a solar panel and a small battery to this DC line and supplement the DC power. With a modest investment, a customer could now use some six tube-lights, four fans, a large LCD TV, laptops, multiple cell phones and tablets.

This innovation does three things at the same time. While it makes blackouts a thing of the past, it encourages customers to use energy-efficient DC appliances. At the same time, it creates a pull for installing decentralized solar power in homes.

In the third example, an NRI materials scientist from Purdue University called V.G. Veeraraghavan framed the challenge to his team of innovators at the Kripya Group of Companies, Chennai. Consumers like government offices, village schools and post offices work only during the day. Can the team deliver solar electricity even to remote places where there is no grid, such as in hilly or tribal areas? Could the team design a solar-powered device that would deliver AC power to the home? During the day, when there is sunlight, schools, post offices, banks and small shops could continue functioning, irrespective of grid power. The team at Kripya designed a micro-inverter, which could be attached to the solar panel. This device could be operational even when the grid power was totally absent. The team is continuing to improve the device that they have already designed.

In the last case, Tata Chemicals framed an industrial question: Can the team save power costs by designing an on-grid system which uses the DC electricity from solar cells on already installed AC devices? A solar module generates DC electric power when illuminated by sunlight. The inverter converts it into AC, which can be used in various applications of energy consumption.

Suitable rooftops were selected from within the factory structures. The array of solar-cell panels generates DC power, which is fed to an inverter, which then converts it into AC power synchronized with internal grid power. The inverter will always maintain the frequency of the solar power higher than the internal grid power so that all generated power can flow to the system. The company expects that the unit will generate power for seven hours per day for all the days in a year. The overall efficiency is expected to be about 17 per cent. The team is currently installing a 150-kilowatt system at the fertilizer factory in Uttar Pradesh, with an appropriate commercial return.

The four examples are distinct innovations, but they share a platform of technological similarity and a journey of seemingly never-ending improvement possibilities.

Coping with opposition

Technophobia and scepticism

Innovators must recognize that there will be opposition to their innovation, but they must listen carefully to the concerns of those who oppose them.

On 28 January 1966, a twenty-one-year-old IIT graduate in the then new subject of electronics and computer science was being interviewed for a job as a trainee systems analyst in a progressive, well-known consumer goods company. At the company's Kolkata office, he was terrified to hear violent shouts from dissident employees and see huge placards opposing computerization. There obviously was a virulent demonstration against computers. To his unbiased mind, computers were being adopted increasingly. In fact, the Indian Railways had just begun a massive programme to install IBM 1401s in the nine zonal and three production units of the railways. To the youngster, computers were the sure harbingers of the future, but even he could not have imagined how transformational they would be. How would this happen in the face of so much opposition, he wondered. I was that young man.

As my career developed with Hindustan Lever, I learned about the fierce opposition to Dalda, a popular brand of hydrogenated vegetable oil. In 1947, Mahatma Gandhi's newspaper, *Harijan*, carried a report that quoted research at the Haffkine Institute, Mumbai, that Vanaspati consumption could result in inferior growth and even blindness. As subsequent events have proven, the idea was quite ridiculous.

Opposition to innovation has always existed. It is called technophobia, which is the fear or dislike of advanced technology or complex devices. In 1675, a group of weavers destroyed machines that replaced their jobs. Around 1811, opposition to machines was led by the Luddites. After the Second World War, the fear of technology grew because of Hiroshima and Nagasaki.

In the 1950s and 1960s, the environmental movement began with investigations into the lead content of petrol. By 1955, DDT had been adopted widely by the World Health Organization to eradicate malaria. Due to Rachel Carson's work on the residual toxicity in plants and animals, DDT was progressively banned in developed countries. DDT continues to be used in emerging markets in the absence of a malaria-control substitute. In 1965, Hubert L. Dreyfus, then a professor at the Massachusetts Institute of Technology, criticized the newly emerging work on artificial intelligence (AI). As subsequent work on AI has shown, this criticism was odd coming from an acknowledged intellectual.

Innovation is slammed by environmentalists and also by legal and bureaucratic action. Uber is an example. It is not a taxi company but a digital platform that facilitates ride-sharing by connecting passengers with taxi drivers. Even an advanced economy like Germany is trying, like King Canute, to hinder its activity. Uber senior officer Emil Michael incited his opposition by making offensive comments about the media, an act that the CEO had to publicly disapprove.

The early history of cars shows how fierce the opposition was, right from 'farmers, horse breeders, blacksmiths, carriage makers, and livery stable operators'. The adoption of computers and cell phones also show a similar pattern. Each innovation, like each human action, produces consequences, some unknown and some expected. Progress necessarily means that while one problem is solved, another may be created.

As protagonists, innovators must recognize and accept that there will be blockers to their innovation, but they must listen carefully to the concerns. Blockers must learn from history that they may delay relevant technologies, but not forever.

Genetically modified seeds

If your child's life could be saved, or if your family could avert hunger, would you oppose experiments in genetic science?

Innovation requires experimenting. Opposition to trials aborts ideas inside the innovation womb. History shows that innovations have faced naysayers, but persistent scientists managed to demonstrate benefits; thus, the naysayers were either persuaded or overcome.

Genetic science is highly emotional, particularly genetically modified organisms (GMOs). In February 2015, a group of British parliamentarians produced a very balanced report on the scientific evidence, precautionary principle, role of non-government organizations (NGOs) and public engagement. As if to muffle the voice of the report, the Scottish government promptly banned all GMOs. GMOs have improved Indian cotton productivity by 60 per cent over the last fifteen years, making India the largest cotton producer in the world. Neighbouring Bangladesh is reported to be achieving breakthroughs by growing the Bt brinjal.

Opposition exists against gene-level trials for human health. *The Economist* reports about a CRISPR (or 'clustered regularly interspaced short palindromic repeats') technology, through which a defect in a human genome can be 'snipped with molecular scissors' and the gap can then be filled with genetic material derived from bacteria. If successful, this technique can provide remedies for many human diseases. But moralists and ethicists are up in arms.

Citizens need exceptional common sense to deal with such issues. Recalling that we *Homo sapiens* have evolved through cross-species tinkering helps according to anthropologist Yuval Noah Harari. Derived from the genus of apes, several human species evolved 2 million years ago and 'acquired pompous titles like neanderthalensis, erectus, soloensis and rudolfensis', he said. These species mated with each other as evidenced through modern DNA data.

Genetic modification is the insertion of DNA from one species into another. According to C.V. Natraj, a former Unilever senior scientist, 'An organism is a GMO when a gene is introduced asexually into another.' Thus *Homo sapiens* (which, incidentally, means 'the wise ones'!) can be regarded as being genetically modified, because cross-species breeding has been a part of our natural evolution! Surely, we are not unnatural creatures, are we?

Bt brinjal and Bt cotton are both created by inserting a naturally occurring gene (not artificially synthesized) from soil bacteria into the plant genome. Our genome mysteriously contains forty genes from bacteria! Yet, cross-species interjection is viewed suspiciously by GM opponents for reasons best known to them. Natraj postulates another mystery, 'Our human body contains billions of small bacteria-like organelles called mitochondria, whose DNA behaves quite independently of our host DNA. In fact we humans harbour more bacteria cells than human cells!'

The message is that the insertion of foreign organisms is not harmful per se, it is nature's way. Citizen referendums to discuss science are bizarre. Highly technical advances should be guided by scientific protocols rather than by political processes. Innovative science can potentially benefit many more people compared to the few opponents.

Incidentally, from an animal point of view, the *Homo sapiens* species is very harmful. They exterminated other species and learnt to dominate the earth. According to Harari, the human brain learned to plan alternative actions as well as the art of collaboration by creating stories and myths—a skill that animals do not possess These created a cognitive revolution, followed by the agricultural revolution from the perspective of *Homo sapiens*—but a disaster for other species!

Developing the story

The innovator must tell and sell the story

Every innovator is a hero to himself. They suffer from the conviction that their innovation is superior to others. They may be right or wrong, only time will tell. But till then, and indeed long thereafter too, the story of what innovation means and how it will benefit others needs to be told. A brilliant innovator who cannot sell his idea suffers a handicap.

Steve Denning worked at the World Bank, and his function was knowledge management.[4] Like many managers, he too believed that analytical was good, anecdotal was bad. All business thinking and training is about being rational, analytical and logical—almost as though, once the logic is clear to someone, he will do the obvious! Yet, every manager knows that is not true. Analysis may excite the mind, but it rarely offers a route to the heart. Not surprisingly, Denning found it an uphill task to persuade his colleagues at the World Bank to accept knowledge management through his logical PowerPoint presentations.

He then told them a 150-word story about a health worker in Zambia. It brought out a human perspective, being packed with emotion, about the state of the problem as perceived by the grass-roots-level worker. He was amazed that his audience at the World Bank was now 'connecting'! Denning learnt the value of the story for sure, but he learnt one more important lesson—that the story or anecdote must be told in a minimalist way, even lacking texture and detail, if need be, in order to attract the attention of managers. This is so because managers in their workplace mindset and attitude have little time and patience for long 'maximalist' stories with a great deal of embellishment and detail.

One of Hollywood's top screenwriting coaches Robert Mckee has a PhD in cinema arts. His students have written, directed and produced award-winning films like *Forrest Gump*, *Erin Brockovich*, *Gandhi*, and many others. Mckee feels that executives can engage listeners on a whole new level if they toss out their PowerPoint slides and learn to tell good stories instead, because 'stories fulfil a profound human need to grasp the patterns of living—not merely as an intellectual exercise, but within a very personal, emotional experience'. In a story, you not only weave a lot of information into the narration but you also rouse your listener's emotions and energy. Storytelling is related to management.

The monomyth of the hero's journey is the common template of the tales of one who goes on an adventure, faces a crisis, wins a victory and then comes home transformed.[5] Consider that Gautam Buddha, Moses, Sri Ramachandra, the Pandavas and Christ are all stated to experience monomyth stories.

Think of the best lessons you have learnt about soft subjects like character, self-esteem and honesty. Almost always, the lesson is associated with an anecdote from your own experiences, is an interaction with somebody you respect or is a story told by another. That is why stories are a very important way of teaching and imparting knowledge, especially about fuzzy and complex subjects. When it comes to ethical and religious studies and subjects such as good character, good citizenry, good social values, and so on, storytelling is effective.

The drama of human emotion is a great preservative for ideas because both the idea and the drama get indelibly etched in your mind—the selflessness of Hanuman, the righteousness of Yudhisthira, Aesop's hare and tortoise, the love between Heer and Ranjha, and so on. The strong connection between learning on the one hand, and anecdotes and stories on the other, is because an idea is united with an emotion.

The development of the photocopier illustrates this point rather well. The dry copier, an essential part of our lives, and whose innovation biography we don't think much about, is instructive.

Developing the dry photocopier

When I began my professional career in the 1960s, Indian firms deployed 'photostat' machines, a wet technology involving liquid chemicals and ink. The principle of this machine was based on photography—an image of the material was placed on a metal stencil and then copies were rolled out by running ink and chemicals over the stencil, much like the letterpress-printing technology. It was cumbersome and slow. The demand for the machines was serviced by photographic and chemical companies like Eastman Kodak and the Haloid Company.

The restless and inspiring atmosphere of the Budapest cafes of the 1930s was a spur to creativity, as evidenced by the story of László Bíró, the inventor of the ball pen, which follows later in this book. The cafe culture spurred another innovation. Hungarian researcher Pál Selényi used to research at the department of applied physics at Budapest University after studying physics and mathematics. Selényi published a paper in a German journal during the 1930s, in which he contemplated dispensing with chemicals and wet processes and proposed that a beam of ions could be used to create a dry image on a rotating drum of insulating material. Thus, an idea crystallized out of a concept in his brain—with no disrespect, an obscure idea from an obscure scientist in an obscure journal!

Chester Carlson, twenty-two years younger than Selényi, was a prolific innovator, who had been fired from Bell Laboratories, New York, for 'his failed business schemes outside of the company'. Many years after its publication, Selényi's paper gripped the mind of the ever-so-curious Carlson. He promptly hired an out-of-work Austrian physicist called Otto Kornei to assist him. Carlson is generally considered the inventor of photocopying.

By 1938, Carlson and Kornei had dry-transferred some lettering from one surface to another, and this was the first prototype of the dry copier technology, which rules our lives today. Carlson got a patent by 1942, but it took another three years for Battelle Memorial Institute to give Carlson some funding.

The leading wet copier company of those times was the Haloid Company. Its CEO, Joseph Wilson, was initially reluctant, but was persuaded to commercialize the dry copier invention. Spurred on by a $100,000 technology grant from the US Army, Haloid secured an inner track for further product development through an exclusive licence for the invention.

A branding exercise of sorts combined two Greek words to launch a new terminology, xerography: *xeros* (dry) and *graphein* (writing). When Haloid directors voted to adopt the term 'xerography', the company's legal department wanted to patent the word. John Hartnett, an advertising executive, prevailed upon the legal department to avoid patenting the word by saying, 'We want people to say the word Xerox.' In due course, the Haloid Company renamed itself Xerox Corporation.

In 1961, about thirty years after Selényi's paper was published, the Xerox 914 was shipped out to customers for the first time. For over half a century now, we have not paused to recall Chester Carlson or Pál Selényi. So much for the footprints innovators leave on the sands of time!

But there is a lesson for the adolescent start-ups.

Corporate Ayurveda for adolescent start-ups

On a visit to China, I noticed that Premier Li Keqiang wanted to focus on 'mass entrepreneurship'. China is buzzing with wild-looking entrepreneurs. However, young Chinese people are found wanting in preparedness, funding and the ability to execute.

India too is in a tizzy, described by one knowledgeable commentator as 'the Mahabharata of Indian Internet unicorns', replete with valuations, action and new developments. Among Indian venture capital circles, growth is no longer the buzzword, profit is. Funding Indian start-ups has slowed down. Around 5000 start-up jobs were lost during the recent months.

Participating in an Ericsson–IIT Delhi event on innovation caused me to reflect. Young people bring much-needed freshness and agility to business. Older folks bring experience and wisdom. Both together can do more than either could do alone. Are there lessons that today's start-ups can learn from yesterday's? There are at least five.

The *first* lesson is the acceptance that start-ups can learn from grown-ups. Yes, Unilever, Siemens, Hershey and Tata started over a century ago, but the constant death and renewal of their body cells has taught these companies lessons about growth and renewal over a century. Companies too have life stages like human beings. Companies may die, but the same may not necessarily be the case with intellectual property. Brooke Bond, Lipton and Pond's don't exist as a company, but their brand is alive.

The *second* lesson is that obstacles are essential to spark innovation. Entrepreneurs overcome obstacles through their sheer passion to innovate in a manner that adds value.

The *third* lesson is that start-ups are best nurtured without their becoming overly conscious of their potential future value. An initial public offer, or an IPO, is the beginning of a new journey, not an end in itself. You don't visit maternity wards or kindergartens to identify future Nobel Prize winners, do you? Remember Jennifer Capriati who was declared a superstar by age thirteen and burnt out within a few years? An analysis of the Westinghouse Science Talent Search data of fifty-two years showed that out of over 2000 finalists across half a century, only twenty made it to the National Academy of Sciences. Psychologist Benjamin Bloom studied world-class musicians, sportsmen and artists to understand how their parents supported them early in life. What did he find? The parents of the most successful kids supported them, but did not drive them as 'tiger moms' would, with exhortations of the child's innate genius.

The *fourth* lesson is that irrespective of how much growth a company achieves, the leadership always strives and struggles to develop a growth mindset among its leaders. A growth mindset is what biologists call 'neoteny', or self-renewal. Stanford psychologist Carol Dweck defines a growth mindset as one where the person wants to learn new things and wants to do new things, not greatly fearful of failure.

Finally, the secret sauce of a grown-up company's long life can be summarized through these four characteristics:

i.   A consistent purpose (Siemens: make real what matters; the Hershey Company: sharing goodness is good for everyone)
ii. Highly focused at the core, but experimental at the edges (Tata Steel and Tata Chemicals focused on their core during the 1960s, while Tata software business founder, P.M. Agarwala, used Tata Sons as his tinkering lab)
iii. A clear identity of who they are and why they are in business
iv. A conservative approach to costs and finance

In a nutshell

Just like how the age from five to fifteen, when the child becomes an adolescent, is very important, the phase when a prototype is adapting to become a marketable model is very crucial. Things can go quite awry for the innovator during this phase: learning from setbacks and failures, endless adaptation, coping with opposition and criticism and, finally, the need to tell a better story to investors and bosses to keep activity fresh and alive.

These constitute the recipe for a company-level Ayurvedic treatment, useful for innovators, entrepreneurs and start-ups.

V

Adolescence: A Product Shaped from This Model

Leaving aside market distortions and infirmities, the full penetration of an innovation can be considered to have been achieved when it is accessible and available for *all* people, everywhere; for example, the electric bulb, the ball pen or footwear. By this definition, it may take an infinite amount of time for an innovation to diffuse fully; a purist would argue that even electricity has not reached all of the 7 billion people on this planet. Consumer marketing companies in India will take decades before they are able to send out their basic products to each of the half-a-million villages in India.

To recapitulate, the idea articulated by the innovator has already been through the phase of a prototype and has been further developed into various versions or models. Now one or more models have to be chosen to become the marketable product—the entrant into the market. A business model needs to be designed and executed, a model which may have its own innovation.

Innovations are not just about products. Whatever the definitions may be, the long journey of an innovation begins with the innovative product making its market debut, complete with pricing and a suitable business-model—akin to the young adolescent who is learning to adapt to the harsh realities of the real world.

At this stage of the journey, the innovator may have become impatient, raring to go to market. The importance of being first has been drilled into the innovator's mind. However 'the system' around the innovator prevails upon him or her to get the product as well as the business model right, to dot I's and cross T's.

This stage of innovation is promoted by certain nourishing practices by the innovator and his or her ecosystem:

- Disciplined business-model development
- Acting fast, but with thoroughness
- Getting lucky along the way

The first challenge faced is a process of disciplined adaptation of a successful model for the product to succeed in the market; the second is the race to be fast and yet be thorough; and the third is grabbing little pieces of luck along the way.

Disciplined business-model development

All companies struggle with balancing discipline and creativity in the innovation process. Both are required, but how much of each, and when and how do you achieve the balance? The organization needs a reliable system to encourage a culture of innovation and support that with a system of capturing ideas. I wish to briefly share how the Tata Group Innovation Forum (TGIF) attempted to address this.

TGIF developed a tool called Innometer in alliance with Prof. Julian Birkinshaw from the London Business School to manage this much-required creative spark. It helps Tata companies to assess the social context in their departments, divisions and companies. Innometer assesses the prevalent process and culture for innovation through surveys, one-to-one interactions and focused group discussions. It develops creative tension at three levels—within the company, by pitching one function or geography against the other; between companies to encourage healthy competition;

and finally, between a company and its global benchmark. The creative tension is palpable because Tata companies are used to being measured for business excellence processes.

To get positive traction out of this tension, Tata developed InnoVerse to identify challenges and provide triggers to employees to generate creative solutions. It is based on the concepts of Prof. Clayton M. Christensen of the Harvard Business School.

The group has derived some early benefits from the Innometer and InnoVerse through the willingness of employees to pitch in with ideas once the explicit and implicit barriers were broken.

Anecdotes and stories about innovations, which went through this 'product to model phase' are surely instructive.

Shale gas

I learn from the story of an innovation miracle that has had great impact in the current century: shale gas through fracking technology. It is certainly not as well known as the iPad or Facebook but has arguably achieved as much economic benefit as these innovations. The shale gas revolution has uplifted the entire global economy, created jobs and given the American economy an extra dose of vim through a sort of economic turnaround during the Obama administration.

On 26 July 2013, an oilman, entrepreneur and philanthropist died in Galveston, Texas, at the age of ninety-four. His father, a Greek immigrant goatherd, used to run a shoeshine shop. Using his own earnings coupled with the family's, the son studied petroleum engineering at Texas A&M University. He stood first in his class. He went on to serve in the Army Corps of Engineers during the Second World War. On his return from the war, he avoided employment with the big oil companies, whom he distrusted. Instead, he ventured out on his own and soon he had drilled thirteen wells on land acquired by him.

He was smart enough to infer from the vast government data that there was a huge reserve of oil and gas in America. By the 1980s, it became an obsession with him to find a technology that releases those reserves of oil and gas trapped in what is called 'shale sedimentary rocks', so termed because they do not have the permeability to release the oil and gas by directly drilling into them.

The traditional way of searching for oil or gas is to locate the place where they may have pooled, through geological traps, in the layers of sedimentary rock that are both porous and permeable. Once found, the pools are easy to produce from by drilling straight into them.

This oilman wondered why he could not access the oil and gas that occurred in tight shales. He developed an innovative way of unlocking vast amounts of natural gas from shale formations.

His innovation looks elegant because he did not innovate de novo, but actually combined two established technologies. He perfected the technique of combining the two technologies: a forty-year-old technology of directional drilling with a twenty-year-old technology of hydraulic fracturing. He found that if he drilled straight down and then turned 90 degrees thousands of feet underground (not at all an easy thing to do), he could run parallel to the shale gas formations. Under high pressure, by injecting water, sand and chemicals, he could fracture the entire shale formation. In this way the trapped gas could be released. It took more than fifteen years to perfect the technique and the oilman had to be doggedly persistent in the face of many naysayers. This process of horizontal drilling and hydraulic fracturing is now common industry practice, known generally as 'fracking'.

Thanks to a widespread adoption of this entrepreneur's innovative genius and dogged persistence, America became the world's largest producer of natural gas.

The name of the man who innovated and accomplished all this was George P. Mitchell. Have you heard of him or know anything about him? His long-time associate Michael Richmond said that Mitchell had the guts to do things that other people would be too scared to do: 'Mitchell did not like to hear "no" so you had to figure out an answer to "how".'

In the late 1990s, Mitchell expressed incredulity at the fact that a few upstarts in Silicon Valley could write a software and sell their company for billions of dollars. His energy company had slogged for decades to build a team of 2000 engineers and vast landholdings. It faced an uncertain future and was worth a fraction of the software companies—quite unfair from his perspective.

Such radical innovations produce unintended consequences. Environmentalists are horrified at the development and rapid adoption of fracking. Documentaries like *GASLAND* tell you that fracking has no legitimate place in modern society. Mitchell himself made the maverick observation that he 'favoured tough regulation that would make fracturing techniques better and safer'.

Persil Power detergent launch

There was much to learn from the disastrous launch of Persil Power in the UK in the early 1990s. Persil Power was a laundry detergent product developed and sold in the mid-1990s by Unilever.

In the early 1990s, Unilever's Persil detergent risked losing its market-leader position as independent tests showed major brands to have had relatively similar results in removing stains. Unilever decided that they needed a product with an edge in stain removal. Persil's main competitor, Ariel, had recently introduced

Ariel Ultra, the first of the 'super compacts'—washing powders equipped with chemical catalysts—which (according to the advertising) cleaned better than ever in spite of using less powder. Given Ariel Ultra's success, Unilever needed a new super-compact Persil line. Thus, Persil Power was conceived.

Unilever's research teams found a manganese-based catalyst that sped up the decomposition of sodium perborate and sodium percarbonate, which act as bleaches in the washing process, thereby increasing the cleaning process noticeably, allowing the use of lower temperatures. Unilever decided that the bleaching agents would be an ideal addition to the product, but had concerns over such a major alteration to the formula of one of their main products.

A high-profile example of changing a major brand was New Coke, with a more direct example being in the late 1980s, when one of Persil's competitors, Daz, introduced a new formula that also cleaned better but caused allergic reactions in a small but noticeable percentage of the population. To this end, Unilever decided to split the catalyst agent, together with some fabric softening agents into a new product, Persil Power. This innovative product brand could be used in addition to normal Persil in varying amounts, depending on how tough the stains were.

In May 1994, Persil Power was launched with a large publicity campaign, but a number of problems soon became apparent. Despite the large publicity campaign, the sales of Persil and Persil Power did not significantly increase because Persil by itself was capable of dealing with most stains. The most serious problem was that after a few washes with Persil Power, clothes first started to lose colour, followed by their structural integrity, ripping easily under any significant stress.

Effectively, washing clothes in Persil Power had the same effect as adding bleach to the clothes. Further testing determined that while the effects weren't apparent on new clothes (on which Unilever had performed most of Persil Power's testing), they could become very quickly apparent on older clothes. The effects were largely determined to be due to Persil Power being a little 'too powerful' in the recommended quantities, and a chemical reaction (which Unilever had not detected) occurring between the catalyst agents and dyes used commonly in clothes.

Considering the embarrassment this episode caused Unilever and the prohibitive cost of redesigning the product, they decided to issue a product recall and then abandon the brand. A number of lawsuits were issued against Unilever by retail chains and consumers, but the vast majority of them were settled outside of court. Afterwards, Unilever was able to refine their main product Persil's formula enough to produce comparable cleaning performance without needing a catalyst. This led to a relaunch of the super-compact format as 'New-Generation Persil'.

It is possible that nobody figured this problem out, or, more likely, figured it out but failed to speak up.

Creating a culture of speaking up

In a company board on which I had once served, a proposal to invest in an African country was discussed over three meetings across a span of over twelve months. There was a powerful, positive logic, but on the flip side, there were severe risks. The analytics were presented persuasively by the CEO—after all, a CEO must seem convinced if he or she is piloting a proposal. There was no table-thumping opposition from the directors, but there were many statements of risks and caution.

The CEO saw these as encouragement, but it was the weak absence of 'no' rather than a clear 'yes'. Finally, it came down to judgement and intuition, and after twelve months of deliberation and investigation, the directors agreed not to pursue the investment opportunity.

A few months later, an editorial about the sorry state of that African country appeared in a respected international paper. I shared the editorial with several directors along with a 'thank God' note. Their individual response stunned me as everyone, including the CEO, said, 'I said so while opposing the proposal.' And I thought, if everyone said so, how come the project was discussed over twelve long months?

The answer, it was postulated by one director, lay in the highly respected image of the board chairman. Directors perceived the chairman to be supportive, so everybody had spoken in such a muted way that their message was lost completely!

History is replete with cases of epic and expensive failures, not because of the lack of intelligence or technique but because of a climate that inhibits the free flow of ideas. Very often, not being able to speak up or bring forward bad news creates a spiral of silence that can result in disasters of massive proportion.

Acting fast, but with thoroughness

Innovators are intense about their idea, and they must be paranoid about speed because 'only the paranoid survive'. This is true, but also not completely true. The survivors in the market are those who acted fast, but with thoroughness. Altair, the first microcomputer in the market, did not survive because Apple entered later with a more sound product and business model. Chux diapers from Johnson & Johnson was an early entrant into the market, but the survivor and future market leader was Pampers from Procter & Gamble.

In these early adolescence years, there are many distractions from speed, announcing who is first and disparaging competitors as they emerge.

Who invented email?

V.A. Shiva Ayyadurai, a Massachusetts-based scientist and Indian-origin entrepreneur, has strong claims to be the inventor of email. One should be understanding of the 'I' factor, which anyway is the first letter among innovators![1] Shiva defends himself against contesting claimants to being the inventor, so it is perfectly understandable that he seeks 'to share his personal story of inventing email . . . and to inspire young people with the larger truth, that innovation can take place anywhere, anytime, by anybody'. His moment of pride came on 16 February 2012, when his 'papers, computer code and artefacts, documenting the invention of email were accepted into the Smithsonian Institution' in Washington. He branded it 'email'.

In 1978, fourteen-year-old Ayyadurai, a research fellow at the University of Medicine and Dentistry of New Jersey, was challenged by Leslie P. Michelson, his supervisor, to translate the conventional paper–electronic communication system. The university was a big campus connected by a wide-area computer network. The computer was in the initial stages of being used in the office environment. Michelson wanted to create an electronic version of the inter-office mail system so that the entire staff of doctors, secretaries and students could communicate faster. Ayyadurai envisioned something simple, something that everyone could use quickly to reliably send and receive digital messages. He embraced the project and began by performing a thorough evaluation of the paper-based mail system, the same one used in offices around the world.

There have been contradictory claims made by several people that they are the inventor of electronic mail. The Special Interest Group for Computing, Information and Society has even indicated that there were electronic-mail systems existing prior to the work done by Ayyadurai, and discounts his work as an invention, arguing that the 'Google Chrome browser cannot be considered an invention as many previous browsers existed'. As early as the 1970s, under an American program called ARPANET, electronic mail had already been invented. The legendary computer scientist Don Knuth has gone on record to say that he used electronic mail in 1975. Bolt, Beranek and Newman (BBN), a subsidiary of the mighty Raytheon, positioned its mascot, Ray Tomlinson, as the inventor of electronic mail. There are many who contributed to significant incremental firsts in the development of electronic mail as we know it today. As a pioneer, Tomlinson has himself said, 'Any single development is stepping on the heels of the previous one and is so closely followed by the next development that most advances are obscured. I think that few individuals will be remembered.'

Ayyadurai built a very easy-to-use, friendly interface, with a rich set of features such as Inbox, Outbox, Drafts, Send, Receive, Attachments and Copy. The first system developed by Ayyadurai contained nearly all the features of today's systems such as Gmail and Yahoo. He named his program EMAIL and acquired a copyright for it in 1982. About 50,000 lines of code were submitted to the US copyright office. Many more email components were subsequently developed and further enhanced this absolutely important, most-used digital mode of communication.

All said and done, the email program developed by Ayyadurai is a hugely impressive accomplishment and has had a transformational impact on the way people communicated electronically.

Whom should history credit with the path-breaking innovations that take place? Should it be the one who conceptualized it and created prototypes, or should it be the one who commercialized the innovation for general use and patented it? In both cases considerable effort and intellect is required. Eventually, the world generally remembers and applauds the one who is the face of the innovation and made it available for public use. The world associates Apple Computers solely with Steve Jobs, but the great contributions of Steve Wozniak are less known. Academics cover the achievements and contributions of Thomas Edison in providing mankind with electricity and the first incandescent light bulb, yet there are many contributions made by Volta, Ampère, Oersted, Ohm, Galvani, Franklin, Maxwell and Faraday that underpin Edison's work.

Here is another case about the fabled American bagels and how they developed.

American bagel machines

Bagels are a tough, round-shaped Jewish preparation.[2] Traditionally, the dough was rolled by hand, converted into rings and then boiled in a kettle. This toughened the product, and it was then baked. The resultant bread-like product had an earthy taste and a very hard crust.

In 1960 the *New York Times* defined a bagel as 'an unsweetened doughnut with rigor mortis'. Until then the bagel was available only in cities with a strong Jewish neighbourhood, best represented by New York. Bagel-making was a skilled trade, passed down from father to son, and its manufacture was restricted to the members of the International Beigel Bakers Union, an organization founded in 1907. Further, to appreciate how conservative this union was, I must note that until the 1950s, the minutes of the union meetings were taken down in Yiddish, not English!

By the 1960s, it became obvious that bagel consumption could exponentially increase if more people could make it in more places in a reliable and consistent manner. Enter Daniel Thompson, a math teacher in California who had a track record of innovating things such as a foldable ping-pong table set on wheels, for which he won a US patent.

Thompson served during the Second World War and then studied industrial arts and mathematics. He was a tinkerer by nature. Sensing the need to automate bagel production, Thompson perfected a fully functional machine that took the labour out of rolling and forming the dough. He set up the Thompson Bagel Machine Manufacturing Corporation in 1961 as the vehicle through which he would deliver his innovative bagel-making technique. While the traditional bagel baker could churn out 120 bagels per hour, Thompson's machine could produce 400 per hour. This innovation democratized bagel-making, with the US producing well over a billion bagels each year.

Thompson died at the age of ninety-four at his residence in Palm Beach, California. His innovation was not product-centric, but was business-model-orientated.

Getting lucky along the way

Whether or not one believes in luck, the innovator needs to get lucky somewhere along the way. Who would have thought that the humble ballpoint pen would get a leg-up because of the Second World War!

The humble ballpoint pen

Pen-maker John Loud got a ballpoint-pen patent in 1888, Anton Shaeffer in 1901, Michael Brown in 1911, while Lorenz presented a prototype ball pen in 1924, but none of them was successful in bringing the product to market—until László Bíró came along in

1938. Apparently simple innovations seem to take a lot of time from their conception to adolescence!

So far I have been likening an innovation to human life: a concept in the brain is like the fertilization of a foetus in the womb, after which the concept develops as a 'life'. The ball pen, which took over half a century from conception to adolescence, secured funding eighty years ago after a historic meeting in Budapest. Ballpoint pen or innovation?

Modern folks would not even think of the ball pen as an innovation. It seems to be an inexpensive writing instrument, not worth too much thought. Yet, the ball pen had to fight for a life. Consider that just fifty years ago, Indian banks would not accept a cheque written with a ballpoint—they could only be written with a fountain pen! Today the ball pen's life is threatened by digitization just as the ball pen itself had threatened the life of its predecessor, the fountain pen.

On 3 March 1938, Andor Goy sat at a cafe with fellow Hungarian László Bíró in Budapest. The latter's hands were smeared with the ink left over from continuous experimentation with his new invention as he 'pitched for funding'. Andor Goy was sceptical because of Bíró's lack of experience in the pen industry. The truth was that the innovation of a writing instrument was led by non-pen people: Lewis Edson Waterman was a salesman who suffered because of his inefficient fountain pen, as did George Safford Parker, a telegraphy instructor.

On 25 April 1938, Bíró secured his patent, a business deal to produce and market his innovation and a bulk order of 30,000 ball pens from the British Royal Air Force. Why the air force? Because the Second World War fighter pilots needed a writing instrument that would not leak! Perhaps this is the reason why Bíró wrote in an introduction to a book about his contribution to the invention of the ball pen: 'The readers of this book should remember that the contents of this book should never forget that what they hold in their hands is a hopelessly biased work. What I recount is the truth, but it is probable that the facts and persons presented are not in reality quite as I describe them.'[3]

László Bíró, a not-so-well-off Jewish writer, grew up in Hungary with multiple interests. The neurons in his brain must have been hyper-charged as he pursued many innovations relentlessly: for instance, a water fountain-pen in which water would flow past a thick ink before touching the paper, a home clothes-washer and an automatic gearbox for his red, twelve-cylinder Bugatti, to name just a few.

Nobel Prize winner and fellow Hungarian Albert von Szent-Györgyi said, 'To be an inventor, a person has to see what everyone else sees, but think what nobody else thinks.' And that was the characteristic of Bíró. How did the concept of a ball pen get fertilized in his brain?

As a journalist, Bíró would often visit the printing press where the rotary printing machines evenly spread the ink over the letters, but the ink would dry up the moment it touched the paper. His fountain pen smudged and spread. Why could his expensive Pelikan pen not replicate the printing press? Sitting on his balcony one day, he watched the children playing down below. Their marbles left a wet trail after emerging from the puddle. Why could that idea not work with a writing instrument? Of course, Bíró thought, he would need an ink dye that would remain a fluid inside the cartridge but dry up as soon as it touched the paper. A distinguished chemistry professor pompously pointed out to him the ludicrousness of the idea by saying, 'There are two kinds of dyes—those that dry quickly and those that dry slowly. How can you make a dye that makes up its own mind?'

It was in the midst of such scepticism that Bíró synthesized a concept in his brain like a foetus in the womb, nurtured it into a prototype as if helping deliver a 'baby' and then perfected it as the prototype grew into its 'childhood'. He faced many obstacles, but he was as committed as a mother would be to her baby.

It is mysterious that a country as small as Hungary has produced thirteen Nobel Prize winners since 1905. How come? The 1961 winner, Georg von Békésy, hazards his guess: 'The years of my life in Switzerland were so calm and settled that I felt no need to fight to survive . . . In Hungary, life was different. It was a continual struggle for just about everything, though this struggle was not one where anybody perished . . . There is a need for such struggles and, throughout history, Hungary has had her fair share . . .'

Socialization promotes the exchange of ideas

The creative environment in Budapest was blooming during the days of Bíró. Cafes have played a strong role in the world's intellectual ferment.[4] The Café de Flore in Paris hosted Jean-Paul Sartre, Albert Camus and Pablo Picasso. London Coffee House hosted Benjamin Franklin and his friends. Budapest had a strong cafe culture—something like Starbucks today—where creative people had a space to chat and doodle for hours.

Historian Brian Cowan describes English coffee houses as 'places where people gathered to drink coffee, learn the news of the day, and meet others to discuss matters of mutual interest'. The absence of alcohol, which used to be present in alehouses, created an atmosphere in which it was possible to engage in a more serious conversation. Historians even associate English coffee houses with the intellectual and cultural developments of the Enlightenment period.

When I lived in Saudi Arabia, I noticed the widespread presence of tea shops. These were places where men could gather for some relaxation, a chat amongst themselves, a continuous supply of tea and a tug of the shisha, a shared pipe for smoking. In the absence of alcoholic bars in the country, these became important places for men to congregate. In fact, my company Lipton had built up a huge and profitable business through the supply of Lipton Yellow Label Tea Bags to this hot tea shop (HTS) segment of the market.

Upon my return to India as the head of Brooke Bond Lipton, I learnt something interesting about Indian HTS's Brooke Bond terminology. In the 1950s, Brooke Bond, the leading Indian tea company, wanted to sell to the lower-income consumer. After reducing packaging and other costs, their sales people made an intuitive observation—lower-income households did not have cooking gas at home, so lighting the kitchen fire meant wood or coal. That meant a lot of effort and cost just to boil some water for tea. Hence, the men went away early in the morning to the nearest HTS to enjoy a ready-made cup of tea. The cost per sip was higher than home-made tea, but it made sense to the consumer. Based on this intuition, Brooke Bond developed their HTS distribution system—and it was to stand them in good stead for decades.

In a nutshell

This phase of innovation is approximate to the human journey from adolescence to young adulthood. From alternative models that were derived from prototypes, and out of all the experiences derived, certain models are chosen to productionize as products. Disciplined adaptation, the care taken to be first in the race and getting lucky along the way are important experiences in this journey to market adulthood.

VI

Young Adult: The Product Competes to Grow

This is the young-adulthood phase of the innovation. At this point, a concept has been articulated as an idea and a prototype has been developed to demonstrate this idea. Several alternative models of the prototype have been tested as best as possible and specific models have been productionized for the market. The product as well as the business model (its target customers, how those customers will be reached, the pricing, delivery of after-sales service) have been designed and are ready for implementation. In biological terms, it corresponds to the young adult, who is ready to start on a career with all the education, grooming and preparation of the past several years.

The emphasis at this stage shifts dramatically from the genius of the individual or the inventor team to systems—process skills, social skills and networking skills.

Process Skills

The product that has been designed must be repeatable in quality and must be delivered to the customer in a reliable way. An American food company has demonstrated how a relatively 'simple' innovation could be industrially engineered and processes perfected, so that these become competitive advantages: McDonald's.

McDonald's

McDonald's is among the largest food companies in the world today, serving 70 million customers every day in 120 countries. The business began in 1940, opened by the two McDonald brothers. It was a drive-in restaurant where consumers could quickly pick up a hamburger and a milkshake for consumption in the car. They used known assembly-line production techniques to deliver tasty hamburgers of a good quality on a consistent basis.

The company soon transformed into a global company (almost joining Coca-Cola as representing Americanism during the Cold War years). I visited and spent a fair amount of time with one franchisee in the mid-1980s.

The big innovation and breakthrough came with the business model. The company had a process manual for every minute operation so that repeatability and consistency were assured. How could a hamburger be delivered, hot and tasty, in a target time from the placement of the order? How could the decor and look of a store be standardized, irrespective of whether the store was in the American Midwest or Beijing? How would the brooms be applied to the floors, how would the cash counters complete their task efficiently, how would the restrooms look clean and practical? Every aspect was industrially engineered and converted into a process manual, and every franchise employee had to attend training programmes at what the company called the Hamburger University!

The company's growth and profitability was evidence of how an apparently simple product could be innovated through its business model, giving the company years of competitive advantage.

Two start-ups: A contrast

I had direct involvement with two start-ups during the latter part of my career. Each one of us dreams of being Steve Jobs or

Elon Musk (with their attendant foibles), but most innovators end up with the weaknesses, not the results, of these great innovators. I have observed and been involved at close range with two start-ups: let's call them MCo and ACo. Neither has so far turned out to be of the 'change the world' variety, but the behaviour and performance of its leaders were different. MCo and ACo had similarities: both were ten years old, attracted similar amount of equity capital, were research-based start-ups and hence were facing the challenge of timing. MCo achieved positive profits within five years, whereas even after ten years, ACo was struggling to achieve positive profits. MCo had an estimated equity value of 5x, while ACo struggled for even a par-equity value. What differences did I observe?

The successful venture, MCo, took a measured approach and spent money cautiously. It crossed the river by feeling the stones, to borrow a Deng Xiaoping expression. The founders were collaborative by nature, they paid themselves modestly, travelled with tight budgets and set their sights on multiplying their long-term share value. They postponed gratification, focused on execution, abhorred debt and were respectful enough of business principles to target positive profit and cash flow.

The ACo leader came through as a slick salesman, full of flamboyance. Even after a few years of missing budgets, the founder could not admit to inadequacy in execution. Company officers were paid competitively and travelled comfortably. The founder periodically looked to the funder for more equity. There was an element of narcissism—which means 'success mine, shortcomings others" type of high betting on the future fruits of research, without planning for a downside. Positive profit and cash flow seemed to lurk around the corner every year, but somehow they did not arrive.

Both of these cases—ACo and MCo—illustrate the importance of business process skills and discipline in the young-adult stage of the innovation.

Social skills

An exposition of social skills in a book on ideas and innovation may seem antithetical to some readers. Notwithstanding such an impression, it is worthwhile to emphasize that it is an essential part of innovation development. Ego trips, the inability to listen and building collaborative relationships are important during this phase of the innovation journey.

Humility vs narcissism

Narcissism is a good thing in small doses. The limits that make it dangerous are undefined. Narcissism can become overpowering to the point of focus on the self rather than the task, and of dreaming of ideas rather than execution.

An idea is merely the mental equivalent of a newborn baby. But the miracle of birth is just the beginning of yet another hazardous journey. Some of the hazards are externally induced but some are genetically encoded; these become a part of the human personality. In management jargon, these are called 'derailers' to connote hazards that can derail an otherwise sound leader. One of the greatest dangers to the continued promise of an idea is the innovator's obsession with himself or his idea. The innovator inadvertently becomes Shiva, the potential destroyer of his own idea through narcissism and inattention to execution.

William Shockley: Brilliant but impossible

I first read about William Shockley when I was an undergraduate engineering student. Along with two other scientists, he had won the Nobel Prize for his work on transistors just a few years before I was deciding what to study further. Of course, I knew nothing about Shockley as a person, but his accomplishments as a scientist were certainly inspiring to me and were an influence in the choice of electronics as my main subject. I learnt more about him later in my life.[1] I was fascinated to read that Shockley, a brilliant scientist, worked at Bell Labs. Along with John Bardeen and Walter Brattain, he worked on introducing non-electrical conducting material with germanium and silicon to establish that it could be a conductor. This laid the basis for the transistor revolution and the chip revolution. Shockley went on to establish his own company in 1953 so as to gain commercial benefits from his invention.

Shockley was as brilliant a scientist as he was a poor manager. He arrived in the Bay Area with every chance of high success. He first rigorously interviewed and assembled eight fine, upcoming scientists. Among them were Robert Noyce, Gordon Moore and Andy Grove, who felt greatly honoured to be recruited by a famous scientist in his company, Shockley Transistor Laboratories, to develop transistor technologies. What they found was that Shockley turned out to be paranoid, contemptuous of subordinates and hugely arrogant. He was a tyrannical boss and such a poor leader that each of the eight people working there walked out of the company. They set up Fairchild Semiconductors and, subsequently, Intel Corporation.

Shockley was pig-headed in many matters and he treated these bright, young people with scant regard. He behaved as though the eight were trying to undermine him. When, in frustration, all eight walked out, he called them 'the traitorous eight'. When he was ousted from his own company, Shockley joined the Stanford faculty. He became interested in distracting race matters, dubbing African-Americans as less intelligent than the Caucasians in the population. He was called a 'Hitlerite' by a reporter, and that remark caused him to get into a libel suit. Shockley won a symbolic amount of $1 as damages, but his reputation was in the mud by the end of the suit.

Shockley died of prostate cancer in 1989, almost completely estranged from all his relatives and friends. His children reportedly learnt of his death through the newspapers.

Bob Kearns: Self-destructive or single-minded

Early in his life, Robert (Bob) Kearns had been a high-school cross-country athlete, an outstanding violinist and a teenage intelligence officer in the Second World War. Kearns had a doctorate in engineering and had taught the subject for eleven years at the Wayne State University in Detroit. He grew up near the giant Ford plant in River Rouge, Michigan, and always thought of the auto company as a place that welcomed anyone with ingenuity.

From 1976, his sole focus in life was to battle the auto giants and reclaim his invention. 'I need the money, but that's not what this is about,' he told *Regardie* magazine in 1990. 'I've spent a lifetime on this. This case isn't just a trial. It's about the meaning of Bob Kearns's life.'

Kearns got his idea on his wedding night in 1953, when a champagne cork struck him in the left eye, which eventually led to a permanent loss of eyesight. The blinking of his eye led him to wonder if he could make windshield wipers that would move at intervals instead of in a constant back-and-forth motion. After years of experiments at home and on his cars, Kearns believed his invention was ready. He applied for patents, mounted his wipers on the 1962 Ford Galaxy and drove to Ford's headquarters to demonstrate his novel product. Later in 1967, he received the first of more than thirty patents for his wipers.

Ford's engineers had been experimenting with vacuum-operated wipers, but Kearns was the first to invent an intermittent wiper with an electric motor. Ford engineers swarmed over his car, at one point sending him out of the workroom, convinced he was activating the wipers with a button in his pocket. At the end of the meeting, Ford told Kearns that they would get back to him. However, Ford did not do so and, in fact, stopped answering his calls. Kearns was left on his own.

In 1969, Ford came out with the first intermittent wiper system in the US, followed within a few years by other automotive manufacturers. In 1976, Kearns's son bought an electric circuit for a Mercedes-Benz intermittent wiper that Kearns took apart, only to discover it was almost identical to what he'd invented. He had a nervous breakdown soon after. Police picked him up in Tennessee and his family checked him into the psychiatric ward in the Montgomery General Hospital. When he came out a few weeks later, his red hair had turned white.

All he wanted, he often said, was the chance to run a factory with his six children and build his wiper motors, along with a later invention for a windshield wiper that was activated automatically by rainfall. In the end, his courtroom battles cost him his job, his marriage and, at times, his mental health.

Kearns filed a suit against Ford for patent infringement in 1978, seeking $141 million in damages (a figure eventually raised to $325 million). In all, he filed lawsuits against twenty-six car manufacturers and other companies. Kearns supported himself with disability pay after his breakdown and by trading in foreign currencies.

By the early 1980s, his wife had had enough. 'It had become an obsession,' recalled his former wife, Phyllis Hall, who lives in Arizona. 'I told him, "I can't stand this life." He said, "This *is* my life."' When their divorce was granted in 1989, Kearns was in the midst of his court case against Ford.

After twelve years of litigation, Ford finally offered to pay Kearns millions of dollars to settle the case. His attorney at the time, William Durkee of Houston, estimated that Kearns could have received at least $50 million from Ford and comparable amounts from other carmakers. Kearns refused the offer.

'He wanted to be a manufacturer and supply that system to the automotive industry,' said Richard L. Aitken, a Washington-based patents lawyer who had worked with Kearns since the 1960s. 'That was the most important thing for him.'

In July 1990, a federal jury ruled that Ford had unintentionally infringed on Kearns's patent and awarded him $10.2 million. After the Ford settlement, Kearns turned his sights on Chrysler. In December 1991, a federal jury ruled that Chrysler had infringed unfairly on his patent. Firing his law firm a week before the damage phase of the trial, Kearns argued his own case and was awarded more than $20 million. Chrysler appealed to the Supreme Court, which ruled that Kearns was entitled to the money, but rejected his argument that Chrysler should be prohibited from using his design.

Having gone through five law firms, an exhausted Kearns was unable to manage his multiple lawsuits on his own. When he missed deadlines for filing papers in his cases against General Motors and German and Japanese auto companies, US District Judge Avern Cohn, who presided over all of Kearns's trials in Detroit, dismissed the remaining cases.

By then, Kearns's patents had expired, having passed the seventeen-year window of ownership then in effect. He bought a house on the Wye River, near Queenstown on the Eastern Shore, and entered an uneasy retirement. From time to time, he would call his children and his attorney and talk about reclaiming his patents.

When Bob Kearns died of cancer at the age of seventy-seven in February 2005, the *Washington Post* described him as an accomplished but frustrated inventor. 'Robert Kearns's battles with the world's automotive giants have come to an end. Kearns devoted decades of his life to fighting Ford Motor Co., Chrysler Corp. and other carmakers in court, trying to gain the credit he thought he deserved as the inventor of the intermittent windshield wiper,' the newspaper reported. Kearns carried on his lonely fight all the way to the Supreme Court, one man against the might of the industrial world and a patent system he believed had let him down.

When he died at seventy-seven of brain cancer complicated by Alzheimer's disease, Kearns had gained some vindication in the form of $30 million in settlements from Ford and Chrysler, but he never got what he had always sought.

Early venture capital: Indian experiences

T. Thomas is one amongst India's finest professional managers and was my chairman at Unilever several decades ago. He was also among India's earliest venture capitalists after his retirement as director from Unilever in 1989. He narrates his experiences in a slim book.[2] Between the two funds of his Indus Venture Capital, a total investment into thirty-two companies have been captured as case-lets under standard heads: what was the innovative/entrepreneurial idea, the actual experience, causes of failure or success, lessons and conclusions and, finally, the exit.

Venture Funds experience a small percentage of success, the majority being failures. The analysis of failures as enumerated by Thomas shows a pattern—a lack of transparency, an inability of the promoter to accept differing opinions, a difficult relationship between the investor and promoter, unresolved conflicts, leading to a loss of trust among co-founders, a misjudgement of dishonesty among promoters and the promoter running a one-man show.

To paraphrase from the opening lines of Leo Tolstoy's *Anna Karenina*, 'All successful new ventures resemble one another, but every unsuccessful new venture is unsuccessful for its own special reason.'

Of course, Steve Jobs was an exception, but every person should not assume that he is a Steve Jobs. Innovation is really creativity plus execution. It is 1 per cent inspiration and 99 per cent perspiration. Innovators often assume that it is the 1 per cent that counts. However, the greatest of great ideas falter in the 99 per cent phase. It is there that the innovator's ego becomes the darkest enemy of the great idea.

Economic historian Omkar Goswami has observed that the difference between early Indian entrepreneurs like Dwarkanath Tagore and Walchand Hirachand on the one hand, and others like Ghanshyam Das Birla on the other, is that the former were 'more interested in the idea of entrepreneurship than the tedium of daily execution'.[3]

Humility and the aptitude for the daily grind are essential to innovation, much more so than most entrepreneurs might like to think.

What it takes to succeed

Many will agree that some narcissism is appropriate among the innovators, but how do we judge how much is appropriate? After all, are we not intuitively careful with an innovator who becomes what Hank Paulson said of ex-Lehman CEO Dick Fuld: 'A person who hears only what he wants to hear'?

But more importantly, what does it take to succeed?

Bill Gross, the founder of many start-ups and the incubator of many others, is passionate about start-ups becoming the biggest drivers in any economy; he is also curious about why some start-ups succeed while others fail. He gathered a huge volume of data available from his own investments and those of others. He ranked each start-up on the basis of five factors: (i) idea, (ii) team, (iii) business model, (iv) funds and (v) timing. Clearly, all five are keys to success, but which is the most important and what is the hierarchy of their importance? Wonder of wonders, his findings show that it is the timing of the innovation that ranked first (42 per cent), followed by the team (32 per cent) and, finally, the idea (28 per cent). Considering the number of passionate innovators who imagine that innovation is the act of thinking of an idea, this is a lesson! The business model (24 per cent) and funding (14 per cent) came fourth and fifth. These results are surprising, they surprised even Bill Gross, who has wide experience on the subject and makes important observations.[4]

The percentage numbers are not worth analysing too much, the sequence is worth deep thought. Innovators who are besotted with 'the big idea' would feel sobered upon reading the relative position of their favourite item. Narcissistic I-orientated innovators would do well to note the second position of 'team/execution'. But the first position of 'timing' is the one that is surprising. Once stated, it is not too difficult to accept. Could it be that some well-regarded innovations like Chotukool, Nano and Swach are all great innovations, which got the timing wrong?

My experience with MCo and ACo validated Bill Gross's view.

The public adulation of start-up founders is universal.[5] Reports in the media paint the popular imagery of the buccaneering entrepreneur with cold facts.

However, there is an alternative story as this extract suggests: '[F]ollowing economist Joseph Schumpeter, start-up heroes revolutionize the economy, reorganize industries and find new ways of doing old things . . . But entrepreneurs who can transform the economy are found in companies of any size . . . Existing UK companies, not start-ups, accounted for 90 per cent of UK productivity growth . . . It is a persistent myth that start-ups exploit a better way of operating than currently.'

Deserve before desiring

'Innovation is 1 per cent inspiration and 99 per cent perspiration . . . and in the 99 per cent phase, the innovator's ego becomes the darkest enemy of the great idea.'

I had read about a company called Balcones, an artisanal Texas distillery whose brilliant founder, Chip Tate, started the company in 2008. In 2013, a cash-strapped Tate invited Gregory Allen to invest. Soon after, differences arose between Tate and Allen and, within nine months, Tate got fired. Start-ups are natural institutions of planned chaos—forgive the tautology. Many ills can be traced to the founder's ego or the investor's greed, both of which are cases of desiring before deserving. And not just in the overseas markets.

Indian-media reports about Housing.com suggest that founder Rahul Yadav, all of twenty-six, the son of a naval officer and an IIT drop-out, is a brilliant man. Yadav started Housing.com in 2012 with angel funding.

The distinctive idea of his company was to provide a one-stop housing solution for potential buyers by unifying all housing players in the market and providing all related activities. Buyers could look for property, compare home-loan rates, sign up with banks for loans and draw up lease rentals. Listed properties would accumulate entries by brokers, sellers, developers, and could be filtered through search specifications such as the number of rooms, lifestyle ratings and child-friendliness index. Quite versatile. But as it turned out, the concept was only 1 per cent of the job.

Within just a few months of getting started, the company raised a second round of funding, then a third and fourth round. The fifth round attracted Softbank, no less, and almost 'validated' the company. By December 2014, the company had over $110 million! By April 2015, the company spent $40 million out of the $110 million on brand-building. Investors wondered about the wisdom of such a huge expenditure. Rahul Yadav's response reportedly included statements like 'dumb', 'intellectually incapable', or 'don't mess with me'.

Writer and commentator Haresh Chawla, a partner in India Value Fund, had a soft corner for young Rahul and offered him wise advice: '[Y]our choice to make . . . [is] between being a great, talented individual player, or being a manager, who has got to be among people, collaborate with them, lead them or be led by them, be able to handle their frustrations, and handle their competence and incompetence. It's an ego-crushing journey where you need to learn to allow people their space to work . . . finding a way to motivate people, to pull them in one direction. A journey of frustration but equally one of triumph when you see the team working together and winning—a journey you share with your team.'

Provocations followed, likely on both sides, and finally, the brilliant founder departed with utterances of expletives and uncivil language. Now one has to watch where it will all lead: $110 million money for 1 per cent inspiration without the 99 per cent perspiration. The founder may have had too huge an ego, but I wonder about the investors! The investors may yet rescue their money, who knows!

Haresh Chawla draws a lesson, 'Check the level of founder shareholdings . . . there appears to be no equilibrium between control and ownership.'

So does entrepreneurial failure or innovation failure have any value? Definitely. Failed, egoistic entrepreneurs may not learn much, but a learning-oriented failed entrepreneur, like the founder of Indiaplaza.com, Kothandaraman Vaitheeswaran, may be of great value. He started India's first e-commerce platform as early as 1999, which lasted till 2011, and is candid about his mistakes.[6] In this field there needs to be a sensitive differentiation between the 'I learnt from my mistakes' type of failures to the 'my way or the highway' type of failures.

Learning to listen

Entrepreneurship is the obverse side of the innovation coin. Entrepreneurs create an institution that should be sustainably beneficial to consumers and stakeholders. The founder's ego is a potential toxin for such an enterprise.

How should we think about ego? An entrepreneur uses polar skills simultaneously: single-minded but also open to others' views; imaginative but also realistic; bold but also keen to conserve cash; decides on gut/instinct but also on analytics. A successful entrepreneur has to be some sort of a genius among *Homo sapiens*. However, a different perspective emerges by not viewing the skills as opposite but by seeing them as complementary. It is similar to one being a loving and a disciplining parent, both at the same time, or a loyal and questioning employee. We are all born with some characteristics or deficiencies, which morph as we grow up. To lead a normal life, our body, physiology and psychology adapt to these characteristics. 'Psychological toxins' is one such characteristic that influences behaviour.

It is possible—indeed, desirable—for entrepreneurs to be sensitive and thick-skinned, to be action-oriented and thoughtful, and to be full of self-confidence yet humble. 'Ego manifests as obduracy—quite different from self-confidence, which embraces a rational flexibility,' says K.K. Narayanan, co-founder of Metahelix Life Sciences, Bangalore.

To be both confident as well as humble, a leader has to listen. One can get training on how to persuade, how to be noticed, and so on. But have you ever come across a course on how to listen? Most likely not. Listening is a skill that cannot be taught! I researched 'listening' for my book.[7] I found that the techniques of teaching hearing-impaired people can teach us how to improve listening through five lessons:

Look people in the eye: Many of us take notes as we listen to people so that we can remember things. Some of us are not fully engaged with the speaker. On the other hand, hearing-impaired people look at the speaker in the eye and make sure that they are fully present in the interaction. They absorb more and retain more.

Don't interrupt: In many entrepreneurial situations, there are simultaneous and multiple conversations. That will never happen with hearing-impaired people. They follow a strict protocol of one person speaking at a time. Consensus and agreement are reached faster, than from a heated and overlapping conversation. In the long-term, slower is faster.

Say in a simple way what you mean: Hearing-impaired people are direct and they communicate with their thoughts and feelings. They tend not to hide behind flowery words. They are economical about the way they communicate. For the same reason, they listen well too.

Ask someone to repeat themselves if you do not understand: Sign language is much more evolved than the spoken word. New signs evolve all the time. Signs used by people from one region may be different from those used by people from another region. Therefore, hearing-impaired people do not hesitate to ask for clarifications if they have not understood something.

Be focused: Hearing-impaired people do not multitask, they concentrate on the interaction on hand. They cut themselves off from distractions. With the advent of smartphones and iPhones, hearing people do the opposite.

These are practical ways for entrepreneurs to listen better and maybe retain humility. They cannot lose by following these, they just might benefit.

Networking Skills

Networking is an important skill, particularly at this phase of an innovation. Networks of people almost always achieve more than isolated individuals who regard themselves as geniuses! This is such a well-known adage, yet the media eulogizes the lone genius to the point that society attributes a whole innovation to a single person!

Network with vendors and allies

One form of networking is with vendors or allies who can contribute with better quality and costs to an innovative project. During all the hype of the Make in India programme, it was reported that the 25,000 bronze cladding parts for the 182-metre memorial statue of Sardar Vallabhbhai Patel were being fabricated at the foundry of Jiangxi Toqine Metal Crafts Corporation Limited, Nanchang, China. The company is obviously competent as its website says that the company is 'a full-service bronze sculpture foundry with twenty-nine years' experience'. The steel frame that is to be contoured is also apparently being procured from China. Sardar Patel made in China!

Network with academia

Another example of networking is that which takes place between business and academia. According to one persuasive report:

'Collaboration between academia and industry is increasingly a critical component of efficient national innovation systems . . . It is the role of public policy to foster such linkages . . . Developing countries face even greater barriers to such alliances . . . Collaboration between universities and industries is critical for skills development, innovation, technology transfers and entrepreneurship.'[8]

India needs more discourse on the industry–academia collaboration. IIT alumnus Shail Kumar has authored a recent book on higher education, in which he has argued the case for better collaboration.[9] The NDA government sponsored the prime minister's fellowship scheme for doctoral research, a welcome public–private partnership initiative.

The truth is that we need a more positive mindset all around. I recall a 1983 industry—academia symposium hosted by the Hindustan Lever Research Foundation.[10] The symposium emphasized the emerging and critical need to grow more pulses, India being a vegetarian and protein-deficient nation. Several advocacy papers were circulated in this regard. However, not much happened for twenty-five years until a supply crisis hit the nation in 2010. I was lucky to leverage my HLL exposure and helped Rallis India start its Grow More Pulses programme in Maharashtra, Madhya Pradesh and Tamil Nadu. Thanks to my observation of the former HLL chairman and, later, the director of Global Research in Unilever, Ashok Ganguly, I have had some exposure to the subject. I also experienced the technology response to the acute shortage of soap-making oils through minor oil development in the 1970s; the development of Tata Swach, the low-cost bacterial/virus water purifier through RHA and nanosilver technology (only 27 paise per litre, gravity-driven). According to the World Economic Forum (2008), India ranks forty-third in terms of industry–academia interaction, compared to China (no. twenty-three), Japan (no. twenty-one), South Korea (no. twelve) and the US (which is ranked first).[11]

There is hidden power if universities would re-engineer themselves to meet grand national challenges like Swachh Bharat, Make in India and Digital India.

In 2015, I attended a 'Research for Resurgence' symposium in Nagpur for fifty university vice-chancellors. Here are some observations: firstly, everyone concurred about lowering the barriers to collaboration; secondly, participants spoke about things like our social attitude, our educational system, the education ministry and how Indian parents raise their kids as though it is an India issue, whereas it is actually a global issue; and thirdly, it took some effort to steer the group towards what they could do when they returned to their offices. Their conclusion to convert their academic team into positive-collaboration-seekers was an energizing but self-evident one. Some research guidance is available in this respect.[12]

Based on a three-year study of the experiences and perspectives of twenty-five research-intensive companies that were involved with over 100 university collaborations, 'Seven Keys to Collaboration Success' reads like a statement on motherhood. However, they are eminently sensible and deserve one's review. Apart from ideas like defining the project's strategic context, sharing the vision with both the university and the company teams, establishing strong communications and review mechanisms, the most striking was 'select boundary-spanning project managers with certain attributes'.

Boundary spanning

This term refers to 'individuals within an innovation or transformation system who have or adopt the role of linking the organization's networks both within and outside'. Boundaries exist within every organization, as do boundaries of the organization with external agencies. Managers inevitably face obstacles to their creativity and dynamism at the boundaries.[13]

Successful organizations nurture innovation managers; they encourage boundary-spanning behaviour and provide support with organizational processes and practices. Boundary spanning includes (i) negotiating with non-group members, (ii) resolving disputes within the group, (iii) obtaining resources, (iv) establishing influence networks and (v) helping followers to deal with the external environment.[14]

Robert Thomas, managing director of research at Accenture, told me, 'The best boundary spanners are intensely curious. They get real pleasure from understanding how others live. Boundary crossers are great at learning new languages and idioms; like consummate travellers, they are forever on the prowl for new places, new food and new ideas. They may not make friends everywhere they go, but they earn respect because they show respect for difference.'

There are fabulous Indian boundary-spanners like Homi Bhabha, Verghese Kurien, C. Subramaniam and Narasimha Rao. The remarkable transformation of the filthy city of Surat by a boundary-spanning IAS officer, S.R. Rao, has been much chronicled.[15] Enormous value is extracted by leaders who can adapt, stretch and operate at the boundaries, thus creatively disturbing the silos into which managers get slotted. Anthropologist and writer Gillian Tett has pointed out that the word 'silo' means 'a tall tower or pit to store corn'. She says silos are a state of mind, a cultural phenomenon. Telling stories about the silo effect can be rewarding. I think of two fantastic silo-busters whom I have encountered.

I learnt from former HLL chairman T. Thomas, who led a path-breaking trajectory from 1974. Squeezed between the rising costs of raw materials and price controls, HLL recorded losses for the first time in its history. Meticulously and patiently, Thomas and his senior officers met secretaries, ministers and even Prime Minister Indira Gandhi, making data-rich presentations. There were setbacks and rebuffs, but they just would not relent. Finally, on 20 September 1974, soap was set free from government control, a turning point in the history of HLL.

The other major advocacy and boundary-spanning act I experienced was the retention of a majority shareholding for Unilever. Although foreign companies like BAT, Metal Box and Dunlop diluted their foreign shareholding as per the Foreign Exchange Regulation Act, or FERA, Thomas was convinced that HLL's management culture would erode if Unilever's shareholding reduced to 40 per cent. It took ten years, spanning two chairmen, hundreds of meetings at various levels and persistence in the face of more than one rejection. The government finally allowed 51 per cent shareholding with the requirement that (i) the company should have 60 per cent of its turnover in the core sector and activities involving sophisticated technology and (ii) the exports should constitute 10 per cent. At that time I led the company's exports.

Tata director Nani Palkhivala also struck me as a fantastic boundary-spanner. I knew him only a bit, but his boundary-spanning attribute was striking. Although he was a brilliant writer and speaker, he told me that he had a stammer and had difficulty in writing due to a cramp! Unbelievable but true! He also narrated how he had accepted the brief for Ms Gandhi's 1975 election appeal, got an interim judgment for her continuation as prime minister, but returned her brief when her government clamped Emergency on the nation. He said, 'I want to be consistent with my lifelong convictions and the values I cherish.' When J.R.D. Tata expressed concern, he offered to step down from the Tata board. Later he served as India's ambassador to the US. What a boundary spanner!

TCS was a Tata start-up established by MIT graduates Ashok Malhotra, Nitin Patel and Lalit Kanodia. Soon, an accomplished and senior MIT engineer, Faqir Chand Kohli, moved from Tata Electric and built TCS, brick by brick. Kohli generously attributes a part of his TCS success to Palkhivala's chairmanship of TCS: 'He had vision and was a good mentor . . . He never interfered in the details . . . and gave full backing to me, especially during times of crisis.'

Boundary-spanning behaviour can and should be taught because a number of challenges await a successful boundary-spanner.[16] First, given that one is working across geographical and functional boundaries, distance is a challenge. Second, time zones could be a challenge. Third, cultural differences and diversity in general could pose a challenge. Fourth and last, the lack of socialization of the people in the team could pose a challenge. Indian innovation and economic development require several boundary spanners. A controversial and local boundary-spanner is Patanjali Ayurved.

An innovative consumer-goods start-up

In the relatively staid fast-moving consumer goods (FMCG) market in India, a boundary spanner has made a huge impact through the superb timing of entry and scaling up: the Patanjali brand has made a significant consumer impact in a very short time. I reflected on what the company must be watchful of to convert such a grand beginning into a sustainable, long-term business.

What are the risks they must guard against? The first is excessive brand extension and distraction. So long as the product range is squarely in the wellness space—I include personal products and ghee in wellness, but not mustard oil, noodles or detergents. Stretching the image into jams, noodles, detergents and cattle feed is neither smart nor image-consistent. Patanjali has a choice to make: create scale by being a minor player in a large number of categories or become a significant player in chosen categories.

The second is the ability to deliver a consistent quality, day after day, year after year. FMCG companies have built systems of quality assurance, safety, food standards and general excellence over decades. This is not rocket science, and delivering results reliably requires a strong systems orientation. Because consumers are hassled with lifestyle pressures, they long for natural, Ayurvedic remedies, which stand pre-sold in the consumers' minds. They trust blindly, and that trust must never be broken in terms of ingredients, quality and freshness. That is a tall order to deliver. Some of the quality complaints on social media are horrendous.

The third is to remain focused on the consumer rather than on the competitor, which can take away focus from the consumer. Patanjali should remember that long-lasting, value-creating consumer companies are rarely controversial entities. They are almost self-effacing because they are always trying to strengthen consumer trust! Advertising or product claims that get struck down by legal courts or by standards councils do not augur well for Patanjali. Suggesting that other edible oils in the country carry carcinogens or add sodium benzoate as a preservative and then claiming that their product is 'chemical-free' is avoidable! How can a detergent be chemical-free? The consumer does not really care whether Patanjali deals a death blow to MNCs or Indian manufacturers.

The fourth is product distribution. 'Herbal-related' brands like VLCC and The Body Shop rely on exclusive stores rather than general trade. Patanjali's ability to get corporate stacking in modern trade is impressive. Currently, Patanjali offers tight retailer margins, but reaches 10 per cent of the retail universe. It is expanding distribution gradually on the strength of consumer pull, but there is a long, long way to go.

The fifth is to dilute the political connections of the business. Consumer research data shows clearly the disapproval of consumers when the Patanjali brand ambassador gets involved with political statements or movements. Changes of government regimes can change fortunes, for example, the availability of bank loans, favourable tax breaks, easing up of investigations into pending quality/tax cases, connection with powerful people, and so on.

The sixth is the Icarus syndrome. If an entire business is constructed on the platform of one brand ambassador, there is inherent risk for life after. With growing success, differences of opinion and compatibility among the stakeholders could crack open. History shows that it is only after commercial god-men die that the putrid remnants of their ashram or empire became visible to the public.

From invention to commercialization

I have been struggling with the nagging question on how long it takes for an invention (i.e., when an original idea has been articulated) to become commercial (i.e., when it is widely sold in the market). The grant of a patent is an intermediate step. Based on limited anecdotal evidence, I had averred that it could be thirty to fifty years.

Joseph Swan obtained the patent for an incandescent bulb as early as 1860, but it took another twenty-three years for Thomas Edison and Swan to launch electric bulbs under the brand name Ediswan. With regard to electric irons, a patent existed in 1882, but it took another fifty-four years before a safe, thermostatically controlled electric iron became a consumer product. Sliced bread was patented in 1912, but it took sixteen more years for Otto Frederick Rohwedder to launch his product in the small town of Chillicothe in Ohio—in 2003, the mayor of Chillicothe commemorated seventy-five years of sliced bread. Swachh Bharat enthusiasts might find new knowledge in learning that although the flush toilet was patented in 1778, it took another seventy years for an English plumber, Thomas Crapper, to market a flushing toilet. It is speculated that in the process, he inadvertently lent his name to the act!

It is erroneously assumed that Procter & Gamble's Pampers marked the launch of the disposable diaper. As early as 1942, twenty-four years before the launch of Pampers, a Swedish company called Paulistrom made the first cloth inserts from tissue. American scientist Marion Donovan prototyped her disposable diaper in 1946 and started selling her product in 1949. Johnson & Johnson launched and marketed Chux, an expensive predecessor of Pampers. So, even a seemingly simple product like diapers took a quarter of a century to get to the market from the prototype stage.

While ideas and patents have a sequential life, they also have a horizontal life as illustrated by rollerballs, which have a rolling ball in contact with some suitably thick material. Although the first patent for ball pens was granted in the 1880s, commercial ball pens emerged only in 1939. Based on the same principle, MUM launched a roll-on deodorant in 1955. In 1964 David Engelbart launched a computer mouse, based again on the rollerball. Ball pen to roll-on deodorant to computer mouse–all sharing technologies!

In December 2015, the UK Energy Research Centre published a 'rapid review' on the time taken for new technologies to reach widespread commercialization. The report visualized innovation to be in two composite phases—first, the 'invention–development–demonstration' phase and, second, the 'market deployment and commercialization' phase. For their review, they considered the facts concerning fourteen innovations. Their findings were: first, that the average time from invention to widespread commercialization was thirty-nine years; second, that there was wide variation among the fourteen innovations, some short and some long, compared to the average; and third, that innovations that replace existing products had shorter timelines (twenty-nine years) compared to those aimed at new markets (forty-two years).

In many cases, the basic scientific principle was established long before a product appeared. For example, the photovoltaic cell using selenium was established in the 1870s, but the silicon-based solar cell appeared in 1954. The cathode ray tube was invented in 1897, but it took several decades before the television set appeared in the market.

Three noteworthy points: first, innovation takes longer than many think, quite unlike the journalistic imagery of 'dramatic and disruptive' innovations. Second, innovations have a life cycle, sometimes short and sometimes long. Although the fruit fly lives fifteen days and the tortoise lives 150 years, their biological cycles are amazingly similar. Third and last, it takes a long time because the inventor cannot understand everything and does not realize the depth of his or her ignorance about what it takes to convert an invention into a product. The human mind is not built to acquire details about every aspect of an innovation.

Humbling, is it not?

In a nutshell

This phase of the life of innovations corresponds to young adulthood in the human-life metaphor. The idea has been presented as a prototype, developed further as a model, refined into a product that is consumer-friendly and is now ready to strike out into the market. The emphasis and keys to success move from technology and product to process-orientation, social skills and networking skills. This is a subtle shift, but a defining one.

VII

Maturity: Challenged to Change

In the preceding chapters, ideas and innovation were humanized by viewing the innovator, his or her work and the application areas which engaged the innovator—through the innovator's eyes. It was in this way that innovation was postulated to have stages of life, much like a human being. The stages of life are not about the product only but also about the innovator as a person. Products do have their own stages of life, popularly called product life-cycle.

In the context of this book, the stages of life are about the goings-on from an innovator's perspective and through the lenses that the innovator is using. This makes it easier to understand why innovators feel proprietary emotions for their innovation and idea, just as a parent experiences possessive emotions for his or her child.

There exist many popular perceptions. For example, some people are more innovative than others, or some fields and applications lend themselves more to innovation than other fields, or that some technologies are inherently more exciting than others. Steve Jobs is reported to have expressed the view, 'It is a disease to think that a really great idea is 90 per cent of the work.' Jobs argued that an idea has no value until it converts into a product manifestation through what he called craftsmanship, which is what stands between a great idea and a great product. The idea turns, twists, mutates and changes all the time as it is being converted into a product. In this process of innovation, novelty and consumer delight are added. The final product may bear little resemblance to the original idea.

Consider, for example, the Arabian peninsula. Before I went to live and work in Arabia in the 1990s, I had already visited the region in the 1970s. Dubai was a nondescript habitation in the middle of the desert. I would never have visited Dubai or Abu Dhabi were it not for the fact that I had work to do there, and also because it enabled me to buy a few odds and ends for the family in import-starved India! When I read about the ruler's visions and ideas, it all seemed fanciful and dreamy. Yet, fifty years later, in 2015, the Dubai International Airport surpassed London's Heathrow with its 78 million passengers. Dubai is a major hub for air travellers. The Dubai we see today is undoubtedly an innovation, but the final product bears no resemblance to the original concept or idea. Anyway, the innovation took fifty years to unravel into its full splendour.

This is also the case with Las Vegas.[1] It was founded in 1905 as a halting place for travellers trying to get to California. There was some freshwater in this area, which was otherwise an arid zone, so railroads and wagon trains found it convenient to halt here. Las Vegas was to developing America what Dubai was to Arabia. Then someone got the idea of converting the city into a gaming centre. So the Las Vegas of today bears no resemblance to the ideas discussed in 1905 or to the early prototypes of its innovation as a city.

The fact is that there are no exciting innovation areas per se. There are exciting problems to be solved, there are exciting innovators and there are exciting outcomes of their work. All the excitement is in the observer's eyes. Because human issues have an evolutionary and civilizational cycle, problems do not evaporate or simply vanish through innovation. The same problems manifest in different forms, almost in defiant persistence. Such problems have economic, behavioural and social dimensions well beyond technology.

Therefore, even though some applications may be considered 'mature areas', they still pose challenges to change. This chapter focuses on the maturity of the challenge and newer ways of overcoming the challenge rather than on the maturity of the product in its life cycles.

The fact is that mature challenges beckon innovators and ideas because aspects of the challenge have remained unsolved for a long time. This insight can be well illustrated through examples of four areas that are thought to be 'old': sanitation, agriculture, hotels and cross-border skills.

Sanitation

Consider innovation in the context of the mundane, unspeakable subject of human waste-disposal and sanitation. The idea of sanitation has provided more and more challenges, both technical and sociological, with every new product form.

A brief history of sanitation

The Indus Valley Civilization existed thirty centuries before Christ and was known for its hydraulic engineering. People then had sanitation devices that were the first of their kind. Individual homes sported the earliest-known system of flush toilets that were connected to a common sewerage system. The 1.5-metre-deep and 1-metre-wide sewers made from bricks were joined together seamlessly. It is amazing that our society, with this pioneering heritage of sanitation, is today the country with the most open defecation (OD) in the world.

Housesteads Roman Fort on Hadrian Wall in Northumberland, England, attests to latrines as the best-preserved feature of sanitation in the Roman civilization, known for innovations in waste management. The waste flushed from latrines flowed via a central channel into a main sewage system and then into a river.

In 1186 AD, in the palace of the Roman emperor in Erfurt, Germany, the fumes from the palace cesspools of sanitary waste caused the floorboards to gradually rot. One day, the floors collapsed and hundreds of the emperor's guests fell into the cesspool, drowning in human excrement!

Around the sixteenth century, early forms of the water closet started to appear. A British nobleman named John Harrington built what he called a 'privy of perfection' for Her Majesty Queen Elizabeth I in 1596. This device washed away the waste but lacked the means to remove the fumes. Around 1920, Mahatma Gandhi had stated that 'sanitation is more important than Independence'.

In 1999, when the *British Medical Journal* polled experts about what they considered the greatest medical advance since 1840, the majority cited public sanitation, ahead of antibiotics and anaesthesia! After so many centuries, the idea of sanitation is alive and vibrant, attracting the best minds in the world to create more innovative product manifestations of an idea for human convenience.

The continuing challenge

The innovation challenge seems to have increased in intensity in spite of centuries of work. Broadly, there are three kinds of sanitation facilities. The upper end of the market is served by flush toilets. The middle end of the market is served by pit latrines. The bottom end means OD. Of the 7 billion people on the planet, about 1 billion defecate in the open. Another 1.5 billion people have no access to flushed toilets. That makes 2.5 billion people completely vulnerable to the ill effects of a poor sanitation system.

One Washington-based company has designed a power plant that could feed off the waste from a small city. The University of West England has showcased a urine-powered fuel cell to charge cell phones overnight. Another team from the University of Colorado has devised a system of concentrating solar power and generating enough heat to kill pathogens in the waste so as to produce biochar as a cooking fuel or fertilizer. The Bill & Melinda Gates Foundation has challenged scientists to reinvent the toilet. Under this programme, an Indian company called Eram Scientific has designed an 'eToilet' that can be self-sustaining: it can generate its own energy and water from the waste and, further, they are interconnected through an electronic network. One fears that such solutions, while being innovative and capable of winning prizes, will be irrelevant to the billions who suffer from the sanitation problem.

Bindeshwar Pathak is famous for founding Sulabh International in 1970. The United Nations Centre for Human Settlements has lauded this great innovation. While this technology is successful, it has not yet dramatically reduced the incidence of OD. I will return to the laudable work of Sulabh later on in this chapter.

The problem is not just one of technology, it also involves social attitudes. When problems like OD persist for long, human beings learn to ignore it. Panchayat leaders should play an important part in convincing people to be hygienic and follow good sanitation practices. There have even been suggestions that families who do not use toilet facilities should not be allowed to apply for government grants.

There are several good organizations, particularly in rural areas, called 'rural sanitary marts', which not only bring about awareness but also construct toilets. They need to be strengthened. Sushanta Sen of the Confederation of Indian Industry managed to assemble over sixty companies in a first-time effort last year. With corporate social responsibility, or CSR, becoming mandatory, much more can be expected of such initiatives.[2]

Neighbouring Bangladesh undertook a massive public programme around the year 2000 involving NGOs, entrepreneurs and a national sanitation strategy. The results may qualify as a revolution; it has shifted people's attitudes to OD. In the demographic and health survey, it was found that only 5 per cent of Bangladeshis defecated in the open, whereas in India, the corresponding number was 57 per cent! Surely, India can do better! There is no shortage of government funds.

Ripe for start-ups and innovation

India has eight 'unicorns', which are start-ups where fund-raising has established a company valuation of $1 billion. Indian entrepreneurs always seek innovative ideas. The most banal human activity offers an opportunity. The Swachh Bharat Mission (SBM) deals with OD. According to its website, SBM has government funding of Rs 1,60,000 crore, has built half-a-million toilets during the year and has generated national awareness. In fact, OD has even attracted international attention!

An enduring and endearing book explains how the human colon (the puborectalis muscle) becomes straight when a person squats as compared to sitting on a commode.[3] This is why the Eastern habit of squatting is effective compared to the Western way of sitting. I consider sitting Western because, to be light-hearted, King Louis IV had his throne made with an inbuilt loo! The very successful Indian film *Piku* elucidated and exemplified this technical position with imaginative dialogues by celebrated actors.

In 2010, an American woman got relief from constipation by simulating the squatting position.[4] The satisfied mother, Jody Edwards, and her enterprising son, Robert, designed a top-of-the-pyramid footstool, that Americans are buying for $25. There are also a wide range of supporting accessories under their brand Squatty Potty, launched in 2012 and positioned as 'the stool for better stools'. Over 2 million Squatty Potty units have been sold since then, with over 11 million YouTube views; the innovation apparently sold very well as a Christmas gift in 2015!

The problem is not about defecation alone. In 2015, San Francisco city officials implemented a new 'pee-proof' paint around the city to combat the persistent problem of public urination.[5] Public-works crews painted ten walls in the city with a special UV-coated, urine-repellent paint. If an offender tries to urinate on a wall coated with the super-hydrophobic paint, the urine, instead of running down the wall, will spray back at the person relieving himself, potentially hitting his clothes or shoes. Public urination has been a chronic issue in San Francisco for a long time. In 2002, the city passed legislation banning public urination and imposing a $50–$100 fine for offenders, but the ban has had little to no impact on the problem.

It was also in 2015 that social entrepreneur Joe Madiath of Gram Vikas spoke at TED Talks about better toilets and hygiene. He has said that it is very fashionable to speak in glowing terms about food in all its forms, but after the food is digested the shit is considered revolting to speak about. An interesting thought!

It is not unreasonable to expect great strides in this realm. Indian business is pregnant with the possibilities of entrepreneurship and start-ups; within the government, the prime minister wants fresh ideas from his bureaucracy, though cynics think this is an oxymoron. My hope is that innovations through Swachh Bharat will attract funding and some 'unicorn' might emerge: this unicorn will be real, whereas the others may be suspect!

Technology and behavioural change

The popular perception seems to be that OD occurs due to a lack of toilets. It is, however, much more complex. The SBM encompasses issues of public health, sociology, psychology, innovation and technology, all rolled into one. The major thrust of the government policies seems to have been to provide toilets to people. Surveys by social institutions suggest that in addition to providing toilet facilities, a positive social and behavioural change is required to be brought about. Hence, technology and innovation need to work with other disciplines of knowledge, sought to be achieved through the social and behavioural-change communication elements of SBM.

Bindeshwar Pathak, a true Gandhian, has demonstrated the value of work at the grass-roots level by setting up Sulabh, an outstanding institution in Delhi, completely devoted to the social, technological and psychological vectors of sanitation.[6]

I was humbled and fascinated by the accomplishments of his institute, including the Sulabh International Museum of Toilets, containing scores of toilet models; and deep, practical insight from action and experimentation rather than mere talk. The visitor is exposed to low-cost ideas, ranging from single pit, double pit, open-to-air and biogas recovery; one can discuss innovation queries like the merits of the squatting vs sitting position, vacuum evacuation (as in planes) and the pee-proof hydrophobic painting of walls!

At a New Delhi conference, minister Venkaiah Naidu warned habitual and deliberate OD offenders. I was intrigued to learn that around 40–50 million people out of 500 million practise OD in spite of having access to toilets. It would seem that they require policing, a bit like how traffic constables check commuters on the road for various offences. Imagine the manpower that would be required—it would be impractical!

I first spoke with Prof. B.S. Das of IIT Kharagpur and then separately with IIT Bombay alumnus Ankit Mehta. They startled me with the possibility that low-cost drone technology can be deployed for policing persistent open-defecators! I felt it was a very innovative idea, if executed sensitively.

Das has prepared a proposal titled 'Monitoring Defecation Using Optical and Near-Infrared Reflectance Spectroscopy'. His proposal essentially considers using 'diffuse reflectance spectroscopy' of the soil to create spectral reflectance data of OD areas. Once a base data set is created, sample data can be captured monthly and overlaid on a geographical information system (GIS) platform or backbone. Comparing the sample data with a library of standard data helps to determine which area to focus on. In this way, a low-cost drone could serve as the airborne platform. Further, this monitoring mechanism can also feed the integrated management information system (MIS) reporting under the Ministry of Statistics and Programme Implementation, which is responsible for SBM's effectiveness monitoring.

Ankit Mehta is the co-founder and CEO of ideaForge, which, he told me, has the capability and expertise to deliver such low-cost drones. These can be used for communication as well as for hyper-spectral data gathering. Thus, the optimization of the drone as a resource is possible. Drones can be deployed under the social and behavioural-change framework of the government's programme and could be used effectively in both individual and community-communication initiatives. Drones could also become effective by their novelty factor. In order to take care of any privacy/shaming concerns, the drones can be fitted only with loud speakers and no-photography payload and will need to have a GPS-based autonomous operation feature.

Ancient wisdom teaches us that the human body is a combination of three 'pipes'. The first is the neuroelectrical wiring from the lower spine to the cranium; the second is the air pipe from the nose through the lungs; and finally, the solids-and-liquids pipe from the mouth through the abdomen, to the point of excretion.

As with all complex subjects, unnecessary controversy is best avoided, for example, highlighting that OD was sanctioned by the 700 BC Baudhayana Sutras.[7] Such a statement is anyway counter-intuitive to the earlier and amazing Harappan toilet system. These were long before modern Western-style water closets started only two centuries ago with the French cabinet de toilette.

The DC water experience

As I thought through the aspect of innovation opportunities in mature arenas like sanitation, I chanced upon a fascinating report about Washington DC. The water authority in the city, called DC Water, was reported to be 'among a handful of utilities who are innovating to save or generate money'.[8] DC Water has 2800 kilometres of sewer pipelines connected to a waste-water treatment plant. The matter flushed down a toilet is transported through these pipes where oils, fats, sediments and objects are screened out and the water is disinfected. The remaining sludge is cooked, sterilized and softened. After that the material is digested in an anaerobic vessel where microorganisms eat the organic matter, producing methane gas for generating electricity. The solids are composted to be used as fertilizers. The cost of the brand-new plant is $470 million and the CEO states that 'it is literally like a start-up because our older business-model was about to fail without fundamental change'.

Social communication and people involvement

For decades, outstanding scientists, politicians and social workers have dwelt on the same theme of public hygiene and sanitation. However, there is a magical diffusion of any message when a prime minister says it. The time has finally come for a concerted and synergistic drive on Swachh Bharat. Any multipronged solution requires a coalescing of administrative, scientific and social forces. This aspect too requires innovation. India has a great opportunity now to focus and innovate in a domain where success can deliver huge benefits.

Newspapers have been carrying reports of the prime minister's exhortation on this subject. He wanted banks and similar institutions to help advance the cause. The railways can play a key role in advancing the prime minister's campaign by carrying the message across the length and breadth of the nation, far more effectively than many other institutions. The question is how?

On behalf of the NInC, chaired by Sam Pitroda, I had made a presentation to the railway board on 7 March 2014. The presentation, titled 'Aavishkar', suggested a joint initiative of the railways, Tata Institute of Social Sciences (TISS) and NInC.

The Indian Railways is a symbol of inclusion and innovation in the country, with multiple stakeholders. It transports over 9 billion people every year, almost 25 million each day. It employs 1.5 million people and its operation covers twenty-four states and three union territories. Five steps were suggested in the NInC presentation. They were: first, the use of the Indian Railways to showcase innovations with the goal to inspire, instigate and provoke stakeholders to contribute ideas; second, to bubble up the ideas by implementing a system to acknowledge and capture them; third, shortlisting the innovation ideas; fourth, executing; and fifth and last, rewarding the innovations.

One important element was to launch a Challenges Worth Solving programme that could feature, very legitimately, public sanitation and OD. To achieve any modicum of success, an information technology backbone would be required to capture and process the thousands of ideas. A software was recommended to create a repository of all the ideas and enable ideas to flow swiftly in the value chain. TCS has already announced its contribution of Rs 100 crore to the prime minister's programme.

Envisioning and implementing such a railway innovation programme needs attitude training and skill impartation. On behalf of NInC, TISS said they would be delighted to partner on such a programme. The NInC offered to join the Indian Railways in the funding of the programme. This would require aggressive advertising and marketing. Creative roughs and demonstration advertisements were presented. These aspects are mentioned to emphasize the point that innovation potential lies across the broad chain, not just at the technology end as is popularly assumed. The same theme was presented separately to India Post, which also has a wide reach and huge capability. India Post could also lead.

Agriculture

Agriculture has been in the public eye throughout 2017, what with farmers' agitations across the whole country, loan waivers and expert commentaries. Ever since humankind decided to settle down into an agricultural way of life—a significant change from the earlier nomadic way of life—the subject of farming and its economics has been around. Undoubtedly, the arena is an old and well-trodden one.

India has over 140 million farm holdings, of which 20 million account for 80 per cent of the area. Imagine a community of these 20 million farmers connected digitally—as change agents, as leaders, as a consumer community. It could be the most thrilling start-up in India and could dominate the national narrative.

There is a report co-authored by Y.S.P. Thorat and myself, which has been circulated to over 1 lakh people interested in agriculture. It was titled *Sarthak Krishi Yojana*.[9] Agriculture is staid compared to magical realms like start-ups.

Sarthak Krishi Yojana expressed the hope that the crisis in agriculture would serve as a wake-up call to governments at all levels. Responses ranged from appreciation to scepticism and even cynicism. A senior journalist observed that we should not rely on the government; one agricultural scientist berated me, 'You should give practical suggestions instead of writing about frameworks.' Another tired opinion was, 'Another yojana will not succeed because there are already 445 yojanas (institutions, schemes, awards, stadiums, airports) bearing the names of a celebrated father, his daughter or his grandson.'

Apathy is a poison for innovation because it leads to low expectations and the 'broken window' syndrome. New York was bedevilled by crime in the 1980s and citizens were cynical about a solution. When Rudolph Giuliani became the mayor in 1994, he was determined to break the apathy. His team invoked the work of two criminologists—James Wilson and George Kelling. 'If a window is broken and left unrepaired, people walking by will conclude that no one cares and that no one is in charge . . . The impetus to engage in a certain type of behaviour comes not from a certain type of person but from a feature of the environment.' The story of how the environment was changed and New York was repaired of crime has been told.[10]

Farming is the nation's broken window—regrettably, not the only broken window! The general public interest in how food gets to the table is low. The Indian farming policy has lacked focus despite the good intentions of the government because there has been no overarching and integrated framework for agricultural development as there had been for industrial development. Political and economic institutions make the difference between a system's success and failure, and systemic change requires a holistic understanding of the problem.

Journalist, editor and commentator T.N. Ninan has written, 'What commentators often ignore is the enormous untapped potential of Indian agriculture . . . Changes in agriculture will directly affect half of the country's workforce . . . The reform of market factors (land, labour, capital and technology) and the improvement of governance standards are central to the question of rapid growth . . . China began under Mao by emphasizing change in the countryside while India sought industrialization . . .'[11] Two questions arise: Does our nation have a holistic approach to farming and agriculture? If not, what could constitute a holistic approach? *Sarthak Krishi Yojana* attempts to address these.

Although agriculture is a state subject, there is political sagacity for both centre and states in strategically (not mendacious subsidies) promoting rural and agriculture. Look at how Gujarat, Madhya Pradesh and Chhattisgarh grew agriculture about 8–10 per cent per annum during the last decade, and how the ruling party got repeatedly re-elected. Writing about Madhya Pradesh, Infosys chair professor Prof. Gulati of the International Council for Research on International Economic Relations, or ICRIER.

stated, 'There are several factors driving agri growth but the most important is leadership and its focus.'

The focus should be on farming, not farmers. The passion of the governments, both at the centre and states, should shift from the yogic state of *kshipta* (unfocused and scattered) to *nirudha* (focused and controlled).

A second rice revolution

Better seeds and cultivation methods can be a game changer. Rice is the highest-produced cereal consumed by human beings. About half of the global population derives its core calorie intake from rice. In Africa, where one in three people depend on rice, the demand is growing at 20 per cent per annum.

Rice is a deeply emotional subject all over Asia. Much rice grows in the rich valleys of the Himalayan river systems such as the Ganga, Brahmaputra, Irrawaddy and Mekong. In India, rice connotes religious and spiritual attributes, for example, *anna daata* (meaning, the 'giver of food') and *anna praasanam* (denoting the first time a baby eats solid food, usually rice). Even Honda and Toyota, which are brand names of cars, mean 'rice fields' in Japanese. Imagine an Indian car with the brand name Chawal!

Asia grows 90 per cent of the world's rice and the per capita consumption of rice is flat. This means that as the productivity of growing rice improves, the Asian population also increases by the same amount.

Rice also consumes a disproportionate amount of water, which is becoming scarce all over the world. So, the world faces a twin challenge with rice: there is a need for higher productivity but using less water. A revolution in rice production can occur by producing better seeds or through better cultivation methods. The combination can deliver fantastic results. For sure, the world is seeking a second rice revolution.

Seeds

In the 1960s, a revolution in seeds occurred through high-yielding varieties, followed by hybrid seeds. Traditional rice is a tall plant with a small grain and lots of body. In a strong breeze, the plant sways and loses the grain. To solve the problem, either the root system could be strengthened or the plant could be made short and stocky. The discovery of a 'short rice' variety in the fields of Taiwan helped produce dwarf rice on a big scale.

In the journey for better seeds, genetic modification arrived as a technique. Rice is the first cereal whose genome sequence has been cracked by science. Genome is a bit like the book of life. This book is in a language that has only four letters, A, G, C and T. Codons are formed by a combination of three of these four letters, each codon determining the building block of proteins. In rice, the genome consists of over 20,000 genes, which means that the code is now known for proteins that are involved in determining all aspects of growth and development. If a scientist wishes to make drought-resistant rice, he or she knows which paragraph has to be modified or rewritten. The same is the case with flood-resistant and salt-resistant rice.

In 2000, *Science* magazine published a paper on 'Golden Rice', a genetically modified rice with beta-carotene (vitamin A) in the polished grain. With subsequent developments, it is possible to increase up to twenty-three times the level of this desirable beta-carotene. Golden rice is now a very potent tool to address the vitamin-A deficiency problem plaguing many parts of the world.

System

Apart from developing new rice varieties and hybrids, the system of growing rice has also attracted innovation. For centuries, the farmer was not too concerned about water or labour intensity. He would plant the seedlings in a nursery. After a few weeks, when the plant was prone to weed attacks, he would flood the nursery with water. After the plant's delicate phase had passed, it would be replanted into the field and tended to for the next several weeks. This process has been developed over centuries and is considered robust, but it consumes a lot of water and is quite labour-intensive.

In 1961, Father Henri de Laulanié, a Jesuit, moved from France to Madagascar, and spent the next thirty-four years of his life working with the farmers there. In 1981, he established an agricultural school. Through his work and some serendipity, the System of Rice Intensification (SRI) was invented. Whenever such a process is established, discipline and order also contribute to the increased yield. The disciplined and orderly approach, along with a reorganization of the way resources (rice plant, soil, water and nutrients) are utilized, contributes greatly to the higher yield.

However, the view that there is higher productivity is contested. A group of scientists believes that it is the additional care bestowed that gives more yield and not science.

It is estimated that over 5 million farmers globally have already adopted SRI cultivation, and in the fifty countries where it has been tried, almost 30–40 per cent of water has been demonstrated to be saved. Due to the better husbanding of resources, the yield per hectare has doubled. SRI has thus become the poster boy for 'more from less' and can justifiably claim to be a 'climate-friendly green rice'.

Imagine the potential of such water-saving rice being marketed as environment-friendly rice. In the consumer atmosphere of a rapidly urbanizing India, this surely holds promise.

Skills

Over the years, like other agencies, Rallis India has deployed agricultural-training institutes (ATIs), somewhat akin to the industrial-training institutes (ITIs) for industrial-skills training. The company experience has pointed to the need for skill training in agriculture, just like unskilled industrial labour is prepared for industrial work in an automated factory. Here are the early lessons: first, an attitudinal and habit shift in the farming technique is required of the farmer; second, SRI involves skills in using mechanized transplanters (unskilled farm labour cannot implement SRI), so an intense reskilling of the farmer is essential; and third, the package of practices must be closely taught by the SRI-promoting agency. The yield increases only with a confluence of all the factors of production.

However, whether with genetically modified seeds or SRI cultivation, the path to paradise is strewn with purgatory and pain. But this is true of all innovation, is it not?

Hotels

Running a hotel and managing their upkeep are both thought to be rather 'old' subjects, surely not so happening in this technology-led world. Hotels too can innovate. This subject arose in a discussion I had with hotel professionals, who argued that their industry does not lend itself much to innovating and that too in a sustainable way. A reputed hotelier called Ian Schrager has a different view.[12] In the 1980s, Schrager pioneered the concept of the boutique hotel—high design, distinct character and an emphasis on chic and social lobbies. Recently, he launched Public, a 370-room luxury hotel on the Lower East Side in Manhattan, New York. He describes it as an 'important and revolutionary idea'—democratizing luxury and making it available to everybody. Schrager has 'got rid of those things that people don't really care about . . . like getting service from a person in a military uniform with golden buttons and epaulets . . . in an obsequious manner'.

On the outskirts of Tokyo, Tsuyoshi Iwasaki runs a factory that manufactures replica food.[13] At this factory, the Iwasaki company recreates the most tasty-looking food items that are, in reality, artificial and tasteless. You know this from the acrid smell of resins and paints that greet you at the factory. Designers from this factory go to restaurants and hotels to watch chefs prepare dishes. They return with an architect's sketch, photographs and notes on textures, colours and consistency. Each dish is individually cast to create a silicon mould into which PVC (polyvinyl chloride) is poured. These are then baked, hand-painted or airbrushed to produce a distinctive appearance. The fake, inedible dish can sell for twenty times the price of the real, edible dish!

Even leaving aside such fancy stuff, five-star hotels can innovate in a practical and consumer-relevant way by producing a magnetic customer experience.

Master switches: I am blind without my spectacles, I keep them under my pillow in the night. I also like the room to be pitch-dark. To find my spectacles in the dark, I usually carry a flashlight that goes under my pillow too. On some trips, I forget to carry it—and boy, what an experience that can be!

Some hotels have a wall-mounted master switch which puts off all the lights. However, after switching that off, the journey from the wall switch to the bed is hazardous. Other hotels keep a bedside master switch. To switch that on in the night, one mostly ends up knocking down the telephone, glass, bottle of water, writing pads and flower vases on the bedside table. Fifteen valuable minutes are required to clear these from the table before which you can then finally flick the master switch off and sleep like a baby. In the morning, you are recharged but your iPad and cell phone remain uncharged! I am sure that hotel engineers have answers and solutions but I wonder why the operating staff is not trained to explain.

When I seek guidance from the attendant who escorts me to my room, I feel like I am being viewed as a strange man with strange requests. After literally shooting in the dark, he or she arranges for housekeeping to help me. Unable to answer, housekeeping summons the electrician. By this time I am well past my bedtime. The lesson is that like with all innovations, customer experience occurs at the junction of departments. Breaking departmental silos leads to a better customer experience!

Temperature: The second innovation opportunity is effective temperature control. Anybody who has sat through a day-long meeting in a top hotel emerges with great empathy for a defrosted, frozen prawn.

The subject is far more complex than designing better AC machines. Both matter: the human traffic load in the room and the activity inside. Recently, *Nature Climate Change* published the following finding: 'Building temperatures are based on a decades-old formula that uses the metabolic rates of men, though many women work in buildings now . . . They follow a thermal comfort model that was developed in the 1960s.'[14] The field is wide open for innovations, is it not?

Air conditioning was invented over a century ago and widely adopted around the time Mahatma Gandhi asked the British to quit India. AC usage has penetrated only 5 per cent of Indian homes, so the Indian market potential is enormous! Surely, hotels and air-conditioner companies can begin an open innovation project to conquer this consumer problem, not just in India but globally.

Electric sockets and key cards: Some hotels provide several sockets at the desk level. But most hotel rooms have so few that you have to bend under a table to disconnect something else and plug in your devices. Heaven knows how many bumps I have suffered on my head.

Any hotel that can give you a key card that still functions after sitting next to a cell phone should win a prize for not making tired guests walk all the way back to the reception to fetch a fresh card.

I think these innovations (or 'improvements', if you insist) are far more consumer-enhancing compared to a fancy enhanced mobile interface or mobile onward reservations. To modify an original statement made in *A Streetcar Named Desire*, 'To be innovative, you have to believe you are innovative.'[15]

Another traditional field is learning across borders. Trade and exchange of know-how across borders is as old as humankind.

Cross-border skills

Foreign direct investment (FDI) is vital to world economic growth as well as for India. Hence, I was delighted to speak at the World Forum for FDI in 2014 in Philadelphia. My subject, however, seemed unusual: the connection between FDI and innovation. Normally, FDI interests economists and financial folks, while innovation interests researchers in business, academic and scientific institutions. As I prepared, I learnt some nuances.

First, why FDI? For fourteen years, A.T. Kearney has published its annual FDI confidence index. The 2014 report shows that global confidence in India ranked seventh, just behind Germany and just ahead of Australia and Singapore, which does not seem bad. However, India was number two in 2012 and number five in 2013, so India has also been slipping. Considering the political and economic events in the country, perhaps this is no surprise. There is a good prospect of a recovery in rankings in the near future. The report argued that a core group of developing economies (China, Brazil and India) continues to enjoy widespread confidence among business leaders. About 80 per cent of respondents showed optimism about the global economy, better than the 50 per cent and 37 per cent in the two years before. Considering that the world FDI inflows are about $1500 billion per year, China is impressive at $120 billion per year, but India is not as impressive, with a much lower $25 billion per year.

Second, FDI is associated with a developed-market company expanding into emerging markets. MNCs with technology, brands and proven business-models expand into new geographies. Using this perspective and using China as the example, a Zhejiang Gongshang University paper averred that FDI did not have much spillover effect on RIC, an acronym for regional innovation capability.[16] The research manifested that increasing domestic R & D inputs, strengthening the innovation capabilities and absorptive capacities in domestic enterprises are determinants to improve the regional innovation capability.

Third, can FDI and knowledge flow from emerging markets to rich markets? India is experiencing an imperfect transformation in R & D /Innovation, known to those in the field, but less than widely known. The accomplishments of the Council of Scientific & Industrial Research, or CSIR, laboratories, the space-related institutions, the agricultural universities and the IITs (all government-sponsored) are only vaguely known to the public. The bright spots of these institutions are cloaked with slightly unfair public cynicism about such institutions. Public imagination is positive when it comes to the private sector and multinational companies' laboratories.

Currently, India has as many as 847 MNC R & D centres, representing about 83 per cent of the top 100 global R & D spenders and half of the top 500 global R & D spenders. India's share in global R & D spending has moved up from the early 2 percentage points several years ago to the early 3 percentage points currently, still a long way to go. Patent registrations in the US from India have dramatically increased, albeit on a small base, from ninety-four in 2000 to 200 in 2010. As R.A. Mashelkar once observed, foreign companies are using Indian brains to research their problems and develop patents for them and are then registering the patents in their own country. Therefore, India can attract FDI on the strength of being a positive hub of innovation. This is different from what Yufen Chen described about China.

Fourth and last, three company cases were presented. Tata Toyo Radiator, a joint venture (JV) formed in 1997 with Japanese partners was intended to absorb technology and make and sell aluminium radiators, initially to manufacturers of passenger cars and small trucks. An early challenge of the JV was to interest other segments, like tractors, big commercial vehicles and diesel-generating set manufacturers, to use aluminium radiators, and that too, under Indian conditions of operations. A domestic engineering capability was established. The local engineers applied the more-efficient aluminium radiator technology to replace copper-brass radiators used in the non-passenger car segments. Further, auto customers sourced the condenser separately from the air-conditioning system. The JV successfully developed the condenser, locally including the concept, thermal design and prototyping, and thus expanding the product range. The Japanese partners were sufficiently impressed with the low-cost innovation capability of the Indian engineers. Very soon the JV established its own research laboratory, which was filing its own patents and developing its own IPR. Some ideas developed in the JV are influencing the Japanese partner overseas.

Reverse FDI occurred with Tata Technologies Limited, an Indian company. Culled from the erstwhile TELCO, the Pune-based company developed distinctive engineering design capabilities relevant to the Indian automotive sector. In 2005, the company acquired a US-based company called Incat, which had exposure to global auto companies through product design as well as product life-cycle (PLM) services. In this case, the FDI involved an Indian company buying abroad. A new and eclectic mix of capabilities was nurtured, different from the original Indian company as well as the acquired company. In 2010 Tata Technologies embarked on a bold programme called eMO, standing for electric mobility. The company debuted a fuel-efficient, environment-friendly, street-legal electric vehicle concept at the Detroit North American International Auto Show in 2013.

Ideas don't have border controls and visas. The third case that generated some interest was the scheme to increase boldness and experimentation in companies. If a team makes a genuine error while doing its innovation work, as distinct from errors due to sloppiness and not following standard operating procedures, then the team should be applauded.

Borrowing an idea from the Regensburg factory of BMW, Tata implemented a scheme called 'Dare to Try' into its global operations, based on exactly this principle. Rallis India 'dared to try' to develop crystalline form of one of its herbicides, but abandoned it before the commercial scale-up due to the risks involved in the process. Subsequent to being recognized within Tata for its honest failure, the enthusiastic team developed an alternate process combining chemistry, equipment technology, kinetics control and catalyst usage to solve the problem.

None of these examples will be widely known in the foreseeable future. However, this is what happens day to day in the myriads of laboratories and institutions in innovation. The fillip to innovation through cross-border openness, ideas and FDI is an interesting feature.

The 'can do' spirit

In 2010, I led a Tata delegation to Israel—a very innovative nation. The visit and impressions created a huge buzz among Tata managers. In Israel, soldiers who are *rosh gadol* (big head) are distinguished from those who are *rosh katan* (little head). Rosh-katan behaviour is shunned because it means interpreting orders as narrowly as possible to avoid taking on responsibility or extra work. Rosh-gadol thinking means following orders but still using your own judgement. It emphasizes execution discipline with improvisation. Rosh gadol connotes a responsible 'can do' attitude.

Bitzua, chutzpah and other such expressions come alive in an eminently readable and inspiring book.[17] The ideas are highly relevant for India, especially at this juncture. 'Bitzua' translates into 'getting things done'. This spirit of 'try it, just do it' is all-pervasive in Israel and has led to the country becoming a top destination for R & D. According to Jewish scholar Leo Rosten, chutzpah is 'gall, brazen nerve, effrontery, presumption plus arrogance'.

Israel is an incredibly innovative nation. It ranks the highest in the world in the per capita number of patents filed. In one case, even a hairdresser had a patent on an algorithm for deducing the right shade of hair. About 22 per cent of the Nobel Prize winners are Jewish; among women who have been awarded the Nobel Prize, 38 per cent are Jewish. These are amazing statistics, considering that the number of Jewish people in the world peaked at 18 million before the Second World War, and nowadays, numbers only about 12 million.

India has some similarities with Israel

Israel is a multicultural society with people of diverse ethnic origins. They are garrulous and argumentative. Particularly during the last sixty years, they have lived with a stark fact: that uncertainty from their neighbours is a certainty. The constant challenge to their very being has made them fiercely proud as they seek self-preservation. They are restless in their quest for economic advancement and social progress; they are highly entrepreneurial. They are competitive to the point of pulling each other down wantonly.

India, too, is multicultural and is an enormously argumentative society; India has to spend money on defence because it faces threats to peace from the neighbourhood, albeit far less than Israel. Indians are a restless people who are incapable of doing repetitive tasks for long; they boast of a long tradition of being entrepreneurial.

Now there are some differences, especially with regard to organizing for innovation. Israel is very good at creating innovation engines, India less so. In Indian companies, there is a tendency to focus on a copious generation of ideas. Academics Vijay Govindarajan and Chris Trimble point out that continuous improvement and operational innovation are best performed by the existing structure, which has been tuned to be an economic performance engine. Ongoing operations have already been programmed to be repeatable and predictable, so continuous improvement ideas are easily absorbed into an operational execution cycle.

A lateral or break-all-the-rules idea requires a differently oriented organization, which they have called the 'innovation engine'. Quite the opposite of the 'performance engine', this innovation engine must encourage challenge, be hugely experimental and accept failures. If a breakthrough idea is pushed from the stage of idea to execution without a special planning phase in between, then the idea loses its zing. With incremental innovation, this risk reduces by the well-tuned planning activity of the performance engine.

As a nation, India does have examples of innovation engines, especially in the public sector. India demonstrated chutzpah during the Green Revolution, when America suspended the PL-480 shipments. Chutzpah was displayed when Verghese Kurien relentlessly pursued the White Revolution, or Milk Revolution. When the US denied India the technology of the supercomputer, as the *Washington Post* wrote, an 'angry India' set out to develop the PARAM supercomputer.

The secret lies in culture, not processes. Companies should examine how to develop more rosh gadol, chutzpah and bitzua by influencing the organizational culture rather than only their process.

In a nutshell

In the popular narrative, there are thought to be exciting innovations, exciting innovation areas and exciting innovators. Those are only in the observer's perception. In reality there is a plethora of contemporary problems which attract innovators, innovative ideas and newer technologies to solve what are thought to be old problems—like sanitation, agriculture and services like hotels.

VIII

Ageing: The Struggle for Relevance

Innovation and entrepreneurship are two sides of the same coin that sit in the same bag as 'change'. Reams of paper have been written and many words have been spoken about the inevitability of change and change being the only constant. Change seems to be like rebirth and recycling in nature; hence, the association in this chapter of ageing with innovation seems most appropriate.

Human ideas and innovations are immortal, they do not age; consider, for example, the long-standing quest to augment stand-alone human strength with power. Individual innovators age and ideas get refreshed. That is why human evolution is full of change and transformation. Like human life, innovations too take their own turns and twists, seemingly getting wiser after each experience. I reflected on the subject of how long it takes for an invention to become an innovation.

Time from invention to innovation

Perovskite is a mineral calcium titanate, discovered in the Ural Mountains in 1839, the same year that Jamsetji Tata was born. Around 168 years later, Perovskite is at the heart of a very exciting solar technology.

I guess that the time lag between scientific invention and commercialization is long. Critics push back because (i) innovators don't straddle the value chain from science to consumption and (ii) timelines are anyway declining with new technologies. Critics point, as an example, to a coin-sized chip, branded BatEye. This chip has been placed on the bat in the cricket Champion's Trophy games; it beams wireless data on the ball angle and bat speed, amplifying the enjoyment for the cricket spectator.

I researched data on the time gap from science to commercialization. In 2009, three scientists from the University of Tokyo, having studied time lags in the solar industry, did not find anything conclusive. In another study in 2011, scientists of the Cambridge Institute of Public Health stated, 'We concluded that the current state of knowledge of time lags . . . is of limited use.'

However, I feel that there are two noteworthy principles: first, that timelines are indeed different from category to category, but their life cycles evolve through a similar pattern; and second, that innovations do not follow a linear pathway.

Life cycle: Technology Life Cycle (TLC) refers to the time and cost of developing technology (R & D) and finding ways to recover those costs (manufacturing and marketing). Studies of TLCs indicate that some technologies like paper, cement and steel have a long lifespan. There are continual variations in technology over a long time, while in others, like electronics, the TLC may be quite short.

A biology metaphor may help. Fruit flies and tortoises have a similar life cycle, but fruit flies live only fifteen days while tortoises live 200 years. Around 1900, agricultural scientist Max Kleiber from the University of California, Davis, established a stunning reality—that the number of heartbeats tends to remain stable within a species, so the bigger the animal the slower its heartbeat. Bigger animals take time to use up their quota. (This fact does not make big animals less or more efficient than small animals).

Without a doubt there are difficulties in judging precisely when the science was invented and when commercialization occurred. In particular, the technology industry thinks in terms of 'platform' vs 'application' technologies—desktop-operating systems like Adobe Postscript, Microsoft Windows and Apple Mac are platform technologies with a long durability. Application innovations derive from exploiting existing scientific knowledge into new market uses, like the Adobe PDF.

Apple impresses because of its consumer innovations; its R & D expenditure at 3.5 per cent of revenue is much smaller than, say, Intel, Alphabet or Microsoft. The iPhone innovation included nine 'enabler' semiconductor technologies like central processors, dynamic random access memory, the liquid crystal and lithium-ion batteries, apart from three 'enhancing' technologies, GPS and AI with voice interface. The science underlying these technologies goes back several decades, and Apple's genius lay in integrating these technologies in a distinctive and consumer-relevant manner.[1] The research on this subject is quite instructive in elaborating this view.

Innovation pathway: Studies recognize the non-linear pathways of some innovations.[2] To quote one report, 'The path from initial idea to market-ready product is uneven, and has many twists and turns.'

The linear model of innovation gained credibility after the Second World War, principally due to the recognition of innovative technologies in defence. The atomic bomb had its origins in nuclear and elementary physics, just like radar technology had its origins in the science of microwave radiation.[3]

Application before science

Iterative technology development which follows a non-linear path has been recognized—for example, innovations about market demands spurred through new insights or innovations where application technology preceded science. An example of innovation through consumer insights is how the video cassette recorder was commercialized.

An example of the application preceding science is how the Wright brothers made a flying machine without knowing much about aerodynamics.

Another example of technology preceding science is the innovative technique associated with Chola-period bronzes.[4] The Cholas reigned for a thousand years from 300 AD in what we refer to as Tamil Nadu these days. Chola bronzes are a showcase of creativity and engineering innovation—they adorn museums all over the world and are showcased as marvels.

The five-step technique is thought to be older than the Chola times, perhaps even a couple of thousand years before Christ. This technique entails first, making a model of the figure to be created; second, a mould; third, the preparation of an admixture of five metals (copper, silver, gold, iron and lead) in a molten state; fourth, the pouring of the metal into the mould; and finally, the finishing and polishing. While metallurgical sciences have been known only in recent centuries, the technologies were developed long before, without an awareness of the science.

Yoga and wellness

Yoga is a remarkable example of practice that long preceded scientific validation. It has rapidly transited from being 'old and traditional' to 'modern and innovative' in terms of the interest of consumers and FMCG companies.

Participating in a senior leadership seminar in 1989 on 'Quality Convenience Foods' at Four Acres in Surrey along with twenty-four Unilever managers, I pored over reams and reams of data over an exciting fortnight. The wellness market included natural, chemical-free, organic, traditional and similar attributes in food and personal-care products. Mentored by Unilever director Iain Anderson, we finally presented our judgements for those times: natural, organic and health foods were a fringe market, selling at premium prices to niche consumers; this niche would develop towards mainstream, but only in an evolutionarily gradual way.

Today, three decades later, I am fascinated to learn that the wellness market has indeed become mainstream.[5] The international media reports that FMCG majors are struggling with low growth compared to emerging health-orientated companies, who have collectively snatched as much as 3 percentage points of market share from the biggies. Kiehl's Cosmetics, based on herbals and an apothecary tradition, is an 1851 New York start-up. But after its acquisition by L'Oreal in 2000, it found its sales had quintupled.

A 2015 Global Wellness Institute study has estimated that wellness is one of the world's largest and fastest-growing industries.[6] The research on their behalf by S.R.I International states that the wellness market size is thrice the size of the pharmaceutical industry—$3.4 trillion vs $1 trillion. In 2015 Nestle announced the appointment of a new global CEO from outside, rather than an insider, the more traditional choice of large companies. Reports state that Nestle must be 'looking at a world in which packaged and processed foods can help treat or alleviate health problems'. French food major Danone paid $12.7 billion for a 1991 start-up named WhiteWave Foods. Something significant is happening in the world out there, something truly mind-blowing!

What about India? The Indian FMCG market in 2015 was estimated to be about $45 billion (Rs 3,00,000 crore), of which the wellness slice has been negligible. Evidence suggests that Indian wellness also is rapidly transitioning towards the mainstream. The Madhya Pradesh government has announced a department of happiness—based on yoga and Ayurveda. The Indian wellness resurgence is led by our traditional concepts: yoga (which relieves impurity of mind) and Ayurveda (which relieves impurity of body). My instinct is that this trend is powered by Indians experiencing the Zeigarnik effect with respect to yoga, Ayurveda and indigenous solutions.

Zeigarnik effect

Bluma Wolfnova Zeigarnik was a Soviet psychologist. In her 1927 research paper, she gave the world what psychologists now refer to by her name: the idea that people remember uncompleted or interrupted tasks better than completed tasks. The trigger for her research was her professor's observation that waiters in restaurants remembered unpaid bills better than fully paid bills. Yoga and Ayurveda have been 'interrupted items' for Indians for centuries.

How so? Indian tradition celebrates the confluence of physical, mental and spiritual harmony as a totality. Around 200 BC, Sri Patanjali enunciated the Yoga Sutras, which embraced what modern-day writers refer to as yoga, Ayurveda and dhyana. This intellectual property has survived in this form for centuries.

During the twentieth century the attribute of a good male physique got a fillip when, in 1905, German bodybuilder Eugene Sandow visited Kolkata. His visit triggered a strong interest in good physique. When I was growing up in Kolkata, any boy interested in bodybuilding was greeted with the remark, '*Kee, Sandow hobey naa kee?* [Do you wish to become a Sandow?]' Initially, I even thought Sandow was a Bengali word, no doubt encouraged by the Bengali-sounding 'oh' in the pronunciation of Sandow. In fact, the Bengalis took to bodybuilding in such a big way that the only Indians to have won the Mr Universe title are Bengalis—Monotosh Roy and Manohar Aich.

In the early 1900s, a young man named Mukunda Lal Ghosh graduated from Scottish Church College, Kolkata, and chose the career of spreading Kriya Yoga. From the age of twenty-seven he lived in California after representing India in the International Congress of Religious Liberals. He became famous as Paramahansa Yogananda, authoring the celebrated book, *The Autobiography of a Yogi*.

Yoga continued its strong global advance when, in 1966, B.K.S. Iyengar, the purist, initiated celebrated violinist Yehudi Menuhin into yoga. Within India, for centuries, yoga and Ayurveda lurked around as the activity of indigenous vaidyas, somewhat second fiddle to the upcoming Western medicine system. The Ayurvedic pharmacies of Kerala have been the flag-bearers in institutionalizing traditional knowledge, particularly the Kottakkal pharmacy pioneered by P.S. Warrier. Companies like Dabur, Zandu and Charak started making and marketing their products from the early 1900s and grew successfully as niche players.

Hatha yoga exponent and Ms Indira Gandhi's yoga teacher, Dhirendra Brahmachari, must get the credit for democratizing yoga among modern Indians through his weekly yoga classes on Doordarshan during the 1970s. Twenty years ago the government created the Department of Indian Systems of Medicine, which was reincarnated in 2014 as a full-fledged ministry with a cabinet-rank minister.

The role of the media in this awakening must also be noted. Bennett Coleman's initiative of publishing 'The Speaking Tree' column in the *Times of India* has certainly brought these traditional Indian thoughts and practices into public consciousness. This publication began in May 1996 as an edit-page column. By early

2010 it had developed into an eight-page broadsheet published every Sunday. For the last two years, 21 June is formally recognized as International Yoga Day, and this has further shone a spotlight on yoga within India as well as internationally.

This history of yoga and Ayurveda is important to understand; yoga, Ayurveda and meditation have had a long-standing but unfulfilled existence in the subcutaneous memory of Indians.

After decades of being a niche market (proprietary medicines, honey and personal products)—the size of the Indian products market is estimated to be Rs 5000 crore—these products have burst forth into the collective consciousness of the Indian consumer. The share of natural and Ayurvedic products in the segment markets has started to rise. Among investment bankers, Ayurvedic brands are becoming hot property for sale and purchase, fetching ten to fifteen times EBIDTA earnings, as illustrated by the transactions of Kesh King and Indulekha. The Zeigarnik effect has started to manifest.

Learning new things: Adopting and discarding

With every new vision, the law of unintended consequences kicks in. When Thomas Edison built the Edison electric light station with his 125-h.p. steam engine, he could not imagine its future environmental impact. When Henry Ford desired to make a personal car accessible to every American, it was difficult for him to visualize the pollution aspects of his ambition. Now, a whole century later, the power and auto industries are completely seized by these issues.

Something that is a technological marvel in one era becomes a threat in another. Wonderful, productivity-promising agricultural advances—fertilizers, pesticides and hybrid or GMOs—are later feared by society. The marvels and benefits of geological coal and seam oils have played out and are then later perceived as a threat to the environment. The benefits and consequent adulation of the internal combustion engine have yielded to curbs, emission norms and deep concerns for the future. The twentieth-century wonders of the vigorous and innovative science of chemistry have faced deep societal distrust, pushing the chemical industry to reinvent itself with green chemistry.

Is it possible that electronics and computers will be regarded as dreaded foes in 2050?

Computers and e-waste

The World Economic Forum places India at number ninety-one in terms of its readiness to transform into a digitalized economy. Notwithstanding this poor ranking, the government's Digital India idea is futuristic and compelling. The plan comprises three components: the creation of a digital infrastructure, delivering services digitally and digital literacy. About 2,50,000 gram panchayats will be connected with a fibre-optic network and every village in the country will be covered through mobile connectivity.

Mary Meeker, a partner at the Kleiner Perkins Caufield & Byers venture-capitalist fund, forecasts that India is set to become the world's second-largest smartphone market in 2017. If all this can be made to work, there is no doubt that India will be transformed. Every citizen would fervently hope for success with our digital aspirations. Digitizing carries the issue of e-waste.

This is particularly ominous for India for two reasons: firstly, our national record of surveillance of laws is patchy, as evidenced by the levels of public hygiene, industrial pollution and haphazard urbanization; secondly, because India is set to adopt digitization more rapidly than many other nation in the coming two decades, our growth of e-waste is likely to top the world charts. Therefore, Digital India must have a component to limit e-waste. How big is the issue of e-waste?

What is e-waste?

E-waste comprises wastes generated from used electronic devices and household appliances, which are not fit for their original intended use and are rather destined for recovery, recycling or disposal. Such wastes encompass a wide range of electrical and electronic devices such as computers, handheld cellular phones, personal stereos and even large household appliances such as refrigerators, air conditioners, etc. E-wastes contain over 1000 different substances, many of which are toxic and potentially hazardous to the environment and human health if not handled in an environmentally sound manner. Here are some facts and extrapolations:

The world already produces e-waste equivalent to seven times the size of the Giza pyramid, roughly 42 million metric tons in just 2014, as per United Nations University. India generated 1.7 million tons of e-waste in 2014, which is roughly 1.3 kilograms of e-waste per capita.

If the penetration of electronics and electrical products in India by 2030 has to grow even to today's average world capita, which leads to e-waste of 6 kilograms per capita, the absolute e-waste generation for India will grow at five times the current level to 9 million tons in 2030!

Another way of estimating e-waste generation for 2030 is taking into consideration the US Environment Protection Agency's estimation of 5–10 per cent global increase in the generation of e-waste each year. Given the low penetration of devices in India, our growth may well be 15 per cent on the smaller base. Approximately 15 million tons of e-waste could be generated by 2030—this is scary.

Impact of e-waste

Developing countries with rapidly growing economies handle e-waste from developed countries as well as from their own internal consumers. Though the Indian Ministry of Environment, Forest and Climate Change has made the import of e-waste illegal, a fair amount of e-waste is still illegally imported into India. Currently, a majority of e-waste handled in India is through the informal sector using rudimentary practices. The informal sector's recycling practices magnify health risks. For example, primary and secondary exposure to toxic metals such as lead results mainly from open-air burning used to retrieve valuable components such as gold. Combustion from burning e-waste creates fine particulate matter, which is linked to pulmonary and cardiovascular disease.

While the health implications of e-waste are difficult to isolate due to the informal working conditions, poverty and poor sanitation, several studies undertaken in Guiyu, a city in south-eastern China, offer insight.[7]

Guiyu is known as the world's largest e-waste recycling site, with its residents exhibiting substantial digestive, neurological, respiratory and bone problems. For example, 80 per cent of Guiyu's children experience respiratory ailments and are especially in danger of lead poisoning. Residents of Guiyu are not the only ones at risk. These chemicals are not biodegradable—they persist in the environment for long periods of time, increasing the risk of exposure.

As per a World Health Organisation study, children are especially vulnerable to the health risks that may result from e-waste exposure and, therefore, need specific protection. As they are still growing, children's intake of air, water and food in proportion to their weight is significantly higher compared to adults, and, with that, the risk of hazardous chemical absorption. 'About 4–5 lakh children between the age of ten to fifteen are observed to be engaged in various e-waste activities, without adequate protection and safeguards in various yards and recycling workshops,' said the ASSOCHAM secretary general while publishing an ASSOCHAM e-Kinetics 2016 study on e-waste.

Further, there are data security implications. I froze while listening to an e-waste operator, who downloaded confidential customer data from a pen drive discarded by a top tech company. Imagine e-waste from the home ministry or the army in the hands of state enemies!

On the positive side, e-waste contains many valuable materials like rare metals, which are well worth recovering, provided one uses green technologies. Just like how green power entered the field of electricity generation as a business, e-waste disposal can be a business, a fact which has been demonstrated by some pioneering companies. The threat of generating e-waste should not diminish the national ambitions in digitizing the country and enabling citizens.

Innovation, entrepreneurship and change

Society is raising an obsessive generation of people for the future, obsessive about the increasing pace of change. In 2002 the American War College invented the phrase 'VUCA', short for 'volatile, uncertain, complex, ambiguous'; since then, there have been a huge number of VUCA conferences.

I was puzzled by two questions about VUCA: first, did our ancestors have a leisurely pace of life (the absence of VUCA) and second, which period in human history experienced the greatest change?

Did our ancestors lead a leisurely life?

As far as the first question is concerned, I experienced great satisfaction out of writing a slim book about how social and technological change was experienced by six generations—my ancestors—dating back to 1823.[8] My source of information was, of course, published books and manuscripts but also the oral delivery of memories by the elders in my family—I had begun my project of inquiry in 1984, though the book saw the light of day almost three decades later.

I concluded that life was very fast-paced for each generation as they felt it, though in retrospect, succeeding generations thought of their ancestors' lives as relaxed, slow-paced and full of leisure time. The truth of the matter is that we judge the pace of change not by the change itself, but rather by the gap between the change on the one hand, and the tools to cope with that change, on the other. It is the gap between change and the tools to cope that determine the feeling of change stress. My intuition is that that gap has more or less remained constant. This is the reason why a slow Internet connection while booking a ticket seems as much of an agony as a long queue at a ticket counter. This is the reason why a delayed response to a text message or email causes as much anxiety or anguish as waiting for the postman's visit in the days of yore. This is why a two-hour delay on the Madurai–Chennai train is as irritating as a half-hour delay on the flight.

It is the gap that matters, not the change!

Which period in history experienced the biggest change?

Thereafter, I set out to inquire about which period in history experienced the maximum change. In the modern era, we are prone to view this question with the starting bias that it is surely the most recent decades and century that has experienced the most blistering pace of change. Attend a few VUCA conferences and we get convinced. As Aristotle noted, when carpenters wish to straighten a warped board, they don't put it in a jig that holds it straight; rather they bend it in a jig that bends it in the opposite direction. I continued my inquiry with the same inspiration. Luckily, I read a fine book titled *Human Race*, authored by Ian Mortimer, a historian who set out to study the same question after watching a New Year's television programme in 1999.[9]

While it is true that most people around the world will intuitively believe that the most recent century, driven by technology, has experienced the most change, a few thinking folks ask the question, 'The most change in what sense? Technology, health, education or social change?' This is a very fair question. It also depends on the perspective that the respondent is coming from.

A top food scientist once told me that the ammonia-synthesis process invented by Fritz Haber and Carl Bosch in the 1900s ushered in the greatest change. Why? Urea enabled the world's burgeoning population to feed itself. On the other hand, when I asked an Indian politician, he argued that the independence of India from the British was the most impactful as it liberated millions from centuries of subjugation. And finally, my dear grandfather told me when I was a child that the beginning of a train service in Tamil Nadu between the big city nearest to our village and Chennai brought about the greatest change. All of them are right, who is to challenge the view expressed?

Connecting and communicating triggers the greatest progress

Reading Ian Mortimer's book gave me the opportunity to reflect on whether a single human idea drove the greatest change, though the manifestation of that idea varied from century to century. I concluded that the single human idea that ushered in the greatest change has been connecting and communicating. The human idea of connecting and communicating has, in my humble opinion, been the overarching driver of innovation and change at the time it manifested itself, irrespective of which century or geography one might study. Think of the power and explosiveness of change initiated by the following innovations in human history.

The oldest that I am able to imagine is when hundreds of thousand years ago, our ancestors learnt to stand and walk on two legs instead of four. Their physiology underwent changes, their brain grew disproportionately and they could see predators and food from a vantage point. Above all, they could move about over larger distances by being on their twos instead of their fours.

Next, I imagine was the change from hunter-gatherer to pastoral farming some 10,000 years ago. It promoted social formations, better connections, communications and the use of the larger brain. The impact of agriculture has been well chronicled, both for its beneficial outcomes as also for its environmental fallouts.

Since I must be brief, I will jump several centuries to the invention of the printing press and the advances in shipbuilding in the first half of the second millennium. Suddenly, common people did not have to rely on the church to state and explain God's word, they could read it on paper in their own homes. They could raise questions and debate what is right and wrong amongst themselves. The improvement in shipbuilding, better deployment of the well-established compass and learning the new art of cartography enabled people to navigate further and also deeper into the forbidding oceans. New lands, new people, new products and new ideas were traded across oceans and lands. The period from 1300–1500 AD was one of breathless change, paving the ground for empires and the Renaissance.

I next refer to the innovation of power in various forms to replace or supplement human effort. Horses, steam engines and internal combustion engines all played a significant role in connecting people across long distances and promoting communication among them.

Electricity generation, transmission and distribution must surely qualify as a powerful subset of connecting and communicating. It is an interesting twist that the long-available solar and wind power have developed a strong and contemporary currency after the perceived and proven excesses of fossil fuel.

Telephony, computers and the Internet must surely bear mention as the new big gorillas in the room when it comes to connecting and communicating. They will, for sure, be followed by robotics and AI.

Seeing what's next

For sure, it would be great for innovation if humans could see what will come next, which is the title of a fascinating book. Among all species, only humans can imagine the future, but the sad fact is that they can only 'imagine' the future, they cannot predict it. Management writers have developed techniques to attempt peering into the future through disciplined techniques.

One such school of thinking is the disruptive innovation school.[10] Several consumer needs require expertise to fulfil, like building a house. If somebody could innovate to allow people to do the task on their own, a 'disruptive innovation' ensues. For example, I had to stand in a queue to buy a railway ticket when I was young, but not any longer, thanks to the disruption of the Internet. In this way of thinking, it is stated that (i) an innovation may have a sustaining or disruptive impact, (ii) the resources to ride that disruptive wave need to be developed as capabilities—what the firm has, how the firm does its work and what the firm wants to do and (iii) which part of the value chain offers the best distinctive differentiation.

In another futuristic manner of thinking, model builders convert possible futures into today's experiences so that current choices can be made.[11] To demonstrate how polluted a future city might be, a forecast of the atmospheric contents in the future was computed and a chemical concoction was captured in a bottle to allow today's consumers to experience it.

Whatever the technique or intuitive judgement, innovators do imagine the future and innovate on the basis of their analysis or judgement. Innovation is the trigger for large-scale change. That is why it is so important to learn new things in this ageing phase of innovation biography, just as ageing humans are advised to play Sudoku, chess or bridge as they advance in years.

Innovation is the trigger for human and societal change

It is in the nature of people and societies to get entrenched in their ways of thinking and behaviour. That is why it is so difficult to change behaviour. When a society is forced by an external force (innovation) to change, then the most significant changes occur in human behaviour. The huge role of innovation, and its inevitability, in the development of human society cannot be adequately emphasized.

An American economist Hyman Minsky researched and wrote about financial markets extensively during his lifetime. However, it was a decade after his death that much interest was reignited in his writings after the subprime mortgage crisis. He argued in 1974, '[T]he financial system swings between robustness and fragility and these swings are an essential part of the process that generates business cycles . . .' The *New York Times* referred to the transition from robustness to fragility as 'the Minsky moment'. Stability is a trigger for instability because stability leads to complacency and boom-bust cycles.

Applied to the field of innovation, I wonder what exactly triggers change in society? It is not the invention that triggers change nor is it the innovative product produced out of the invention. The trigger for large-scale societal change is the widespread adoption of the innovation by consumers. Consumers will change when they are jolted out of their zone of·complacence and feel compelled to adapt.

So in closing this chapter and this book, I must reaffirm that while it appears that innovations have a life like a human being, the act of innovation is timeless and reinvents itself. Innovation is life and life is innovation. Both go through the cycle of birth and death but do not vanish, they get recycled.

Epilogue

As I draw to the end of writing this book, I reflect on what it takes to have an idea and innovate. I reckon it is like being a child forever. In biology they term this as 'neoteny', always fresh and young!

There are lots of significant books by wise people, but could I try to figure out an idea, an acronym that the reader can remember easily? I can strengthen that acronym with a real story to improve its recall value. The acronym CHILD flashed through my mind—standing for curiosity, humility, intuition, learning and determination. CHILD as an acronym also seemed appropriate to the biographical context in which I am writing this book.

Famous people abound in the pages of magazines, speeches and books. They are legends in innovation and entrepreneurship—Bill Gates, Steve Jobs or our own Bahls and Bansals. The speeches, writings and books contain truly inspiring stories and anecdotes, both of which are sought by aspiring innovators and entrepreneurs. They are pure inspiration. The founder's vision, the obstacles experienced, the sheer passion and determination and the chutzpah appear exemplary, and are, for sure, contagious.

However, inspiring stories do not always offer instruction. Innovators and entrepreneurs need both inspiration as well as instruction. Instruction happens when the devils of the mind are revealed.

Devils of the mind

So whom does the innovator or entrepreneur turn to for instruction? They turn to stories of ordinary people, not geniuses or icons.

Instruction occurs when there is no glamour and no romantic hues. Instruction happens when we learn from people whom we can touch and appreciate. Instruction happens when their daily grind appears visible, when their natural and human anxieties and uncertainties are palpable and when their psychological throes of self-doubt and hesitation are revealed. I call these the devils of the mind.

To truly understand innovation and entrepreneurship, I think there must be a greater awareness of the devils of the mind. What goes on inside the mind of the entrepreneur? What are the attitudes that shape the person? What foolhardiness enables them to cope with the devils of the mind?

Indeed, there are books that offer to explain the qualities of the mind and heart. Too many attributes and attitudes are mentioned. Keeping in mind the biographical thesis of this book, I commend the acronym CHILD for its simplicity and memorability.

My emphasis on a childlike mindset is illustrated through an example of innovation and entrepreneurship by an ordinary person—a person like us (PLU).

Yash and Satyen Sanghavi

Here is the story of Yash Sanghavi and his son, Satyen. Have you heard of them, or their story? Not really, and that is understandable. They are not famous, they are PLU.

Now a sprightly sixty-year-old, Yash still demonstrates the Kapol bania (a traditional trading community from Saurashtra) instinct for commerce and business. He lost his father when he was quite young. His father and the family were in the engineering business; the family business used to employ 400 people in the activity of manufacturing. Thanks to the Datta Samant wave of labour unrest in Mumbai during the 1970s, the family business closed down by the early 1980s.

After he lost his father, Yash separated from the family business with the princely sum of Rs 30,000—not small money, considering it was in 1984—but certainly not big money either. For Yash it was a near-death experience: What was he to do with all that money and absolutely no knowledge of engineering?

However, this very fear caused him to reflect and take risks. 'When you have little to lose, you feel encouraged to take a risk,' said Yash Sanghavi rather enigmatically.

'A Gujarati Kapol bania is a natural trader,' Yash said. 'For some time I had a nagging doubt whether the family foray into manufactured and engineered products was a smart move. After separating from the family business, I redirected my resources, time and energy back into the natural vocation of a bania—trading and indenting, choosing chemicals as my field of play.'

Yash Sanghavi had an unobtrusive personality, not at all incisive. His outward behaviour made him appear as a natural connector with people. He built a network of professional contacts with purchasing managers in important chemical-oriented companies like Boots, Merck, J&J, Britannia and even overseas companies like Nishi Iwai. Within a decade he tasted some success through trading in laboratory chemicals. Gradually, he felt the need to connect with international chemical trading companies. He started travelling abroad; he was the first in his family to venture abroad in 1982–83!

As it happens with many middle-class families, Yash was modestly rooted in religion and philosophy. He was a voracious reader of business, philosophy and current affairs, driven by a sense of curiosity. He observed that everyday life offered so much from which you could learn, both from your own mistakes and from other people's as well. Entrepreneurship is about learning all the time.

As is the social custom in Indian families, occasionally indigent relatives and friends would request financial help from the better-off relations. Since Yash had tasted modest success in business, a reassured Yash entertained requests for help as far as he could manage. This built up a strong social network. Later in life he felt that this strong network enabled him to overcome the economically challenging days.

Bolstered with some business savings to fall back upon (much of which he reinvested into real estate) and a determination to be a value-adding chemicals indenter-supplier to chemical companies, he became comfortable. Soon after, however, he sensed a tsunami from economic liberalization. Import tariffs which used to be 240 per cent started to decline steeply and sharply. Foreign suppliers could set up their own offices in India more easily. Contrary to a defensive strategy, Yash figured that in this situation he should strengthen his networks, not abandon them in despair.

A new opportunity started to take shape in the mind of Yash. He was friendly with K.K. Sharma, later vice chairman of Lupin. Their friendship was further bolstered through playing squash. Moreover, while working on his customers and projects, Yash had developed good business relations with Kyowa Hokko through the marketing head in Asia, Raymond Goh, who repeated his view, 'Biologics is important and vital for future medicine.' Sharma who had been in the field of medicine for over twenty-five years, supported Goh's belief. Sharma would say, 'Biologics is a natural process. Even an allopathy manufacturer has to revert to biologics in the future'. The vision of this statement strengthened Yash's resolve to strengthen relations with Kyowa Hokko, a Japanese company that specialized in the niche biotech area.

Culturing, fermentation and creating natural chemicals started to interest a rather curious Yash. On one of his outward travels to Japan, Yash visited Kyowa Hokko, and found him to be very knowledgeable and savvy in what was then the small and unknown area of biotech. The more Yash learnt, the less he seemed to know, and this gap of knowledge caused his fascination with the subject to rapidly increase. In this process he stumbled on to the possibilities enabled by stem cells. Later he would connect with a South Korean company called Sewon Cellontech, who agreed to license their technology to him.

Although Yash was cautious about straying from his comfort zone of trading, his curiosity about biotech and stem cells spun out in a fanciful dream. Yash was unsure what the dream was. All he knew for sure was that it was a dream: a bit unreal though enjoyable, but a dream that may be worth chasing. If he was to do something in biotech, whatever that something would be, he was sure that he needed a family member with technical expertise. He figured that he also needed a stronger network of connections to shape his dreams.

His son, Satyen, was a reasonable student when he finished school. In deciding what degree course to study, the young man appeared like a solution looking for a problem. Luck strikes the prepared, it is said—in the form of a brochure that arrived in the mail from the D.Y. Patil Institute, for a degree programme in biotech. It was not a subject much in demand, so Satyen applied and soon acquired a BTech degree in biotech. In a further denouement of serendipity, Satyen learnt of an MSc programme in stem cell technology in UK's Nottingham University. Now who would think of a plan to study stem cell technology at an expensive institution abroad?

For Yash the building blocks were coming together, a bit like Lego pieces would for an enthusiastic child. He was personally curious; his son could acquire an MSc degree and return to India, so Yash would have his technical expert from the family. Yash could assemble biotech experts from D.Y. Patil (there were no stem cell experts then), the college where his son studied. Could a lecturer on plant tissue culture reorient himself to understand human cell culture? Enter Vinayak Kedage, a PhD from Kolhapur!

Yash was ready to foray into an esoteric arena, very far removed from his Kapol bania roots and heritage. In 2007, Regenerative Medical Services Private Limited (RMS) was born.

On the positive side of the resources, his son would return with an MSc degree, the lecturer on plant tissue culture would relearn human cell biology, Yash had land in Lonavala, and enough money in the bank to get started without other promoters. On the negative side, he had no awareness of what business model to adopt, the regulations around biotechnology and how to spend capital sensibly.

As the late C.K. Prahalad would point out to management leaders, growth requires that ambition should exceed the resources. Yash Sanghavi was a walking example of this magical formula.

RMS describes itself as 'the leading biotech company, focused on the delivery of the most advanced cell therapy treatments'. Using the brand name Regrow, the company aimed to be 'the first in the world to offer bone and cartilage cell therapies'.

Without getting into the technicalities, cell therapy means injecting cellular material into the patient to replace diseased or dysfunctional cells. Cell-based therapeutics and stem cell bioprocessing are intended to repair bone and cartilage cells naturally. Suitable cells would be extracted from the patient, brought to Lonavala, where they would be multiplied, and then sent back for injection into the suffering patient. In technical terms, this is referred to as 'autologous'. Surely that says something big about innovation and entrepreneurship. Develop the cell multiplication technology in India, may be the first in the world, clear all regulatory requirements, and commercialize the product and service!

The company has the father, Yash, as the managing director. His son, Satyen, is the chief scientific officer. Kedage is the lab director. There are forty-seven employees at the Lonavala unit, some of whom are highly qualified or trained in medicine, clean laboratory practices, cell biology and biotechnology. Interestingly, the employees come through as 'ordinary' people, not glint-eyed geniuses; they look like people who are committed, who are learning and growing with 'their' company. There are more than 200 employees in the company, including over 100 sales professionals

Start-up rather than upstart

Unlike many other founders, this father and son did not speak about exceptional growth, valuations, funding and massive cash losses. They did not come through as cocky people at all. They wanted to control costs, focus on distinctive customer benefits, persuade orthopaedic doctors and build an ecosystem of value-adding business partners and employees—all the classical business lessons, applied in a crazy world of alternative models. If they continue on a sensible path, the father and son should stand a good chance of making it big. So where does CHILD feature?

The whole idea was born out of the insatiable *Curiosity* of Yash.

The leaders are visibly *Humble* because they speak about their origins and knowledge base without embarrassment.

Their key decisions—whether to enter the field, how to make a world-class laboratory, whom to hire—were all based on *Intuition* after all the tools of analysis had been explored.

They are proud that they go to sleep each night with a slight improvement due to on-the-job *Learning*.

They admit that they do not have a precise road map for the journey ahead, but their compass shows them that they are moving in the right direction. So they move with *Determination*.

Is this business of growing cells and adopting new types of therapy really so new and innovative? I think so and I am convinced now after my visit to the RMS laboratory. Why?

I chanced upon a report about a new research paper from the University of Florida.[1] Scientists have discovered techniques to make heart cells regenerate, an advance that could lead to new therapies to treat cardiac diseases. Normally, our heart genes have a lock device that tells the heart genes not to grow new parts. However, in the sea anemone there is no lock device. By studying the sea anemone, scientists think they may be able to unlock the human heart gene to produce new cells. If they are successful, our new cells can be grown and used to regenerate our heart muscles.

RMS has learnt how to do so for bone and cartilage—quite impressively, at least to a layperson like myself!

How do you measure the impact of an innovation?

Before closing this book, I will bring up an important question—how do you assess the *impact* of an innovation? Consider these four innovations and assess which one has made the greatest impact on mankind—the invention of anaesthesia, the synthesis of urea fertilizer, the discovery of penicillin and the creation of the Internet/email. The answer depends on how significantly the innovation has been adopted and how much it has changed people's lives. Popular reportage considers metrics like R & D staff employed, technology expenditure, revenue from new products and the number of patents obtained.

Publicity amplifies the 'wow' factor through breathless accounts of individual genius. Do you know of the genius who invented the Internet/email? You don't, because there is a lineage from Leonard Kleinrock (1961), Ray Tomlinson (1971), Vincent Cerf (1973) to Tim Berners-Lee (1990). Had it not been for the sequential adaptations through this lineage, consumer adoption would not have occurred, thus creating human impact. Financial impact is important, but human impact is far more important. How many people are aware of the innovation and have experienced it? How many have adopted it? How has it changed the lives of those who have adopted?

Brands measure their impact using similar metrics. Brand share is brand adoption multiplied by usage. The same principle can be used to measure innovation impact. The theories of Pierre Bourdieu (1930–2002), a French social scientist, are also useful when thinking about innovation impact. Bourdieu had said that power in politics arises from social capital and habitus.

Innovation impact is the multiplication of innovation capital, innovation habitus and innovation adoption.

Innovation capital derives from: (i) intellect (technology, patents, trademarks); (ii) articulation (the idea being heard and understood); (iii) culture (initial struggles, obstacles); and (iv) symbolism (recognition and awards). Several Silicon Valley innovators who graduated from or dropped out of top schools were great articulators of their ideas, acquired visibility by mingling and appearing with the high and mighty at Davos-type events and acquired great valuations. These factors contribute to accumulating innovation capital.

Innovation habitus is the knowledge of how to negotiate the world and how to develop stories, mannerisms, opinions and style. Recall how Uber was widely reported to have been dreamt up after someone failed to get a cab, how Airbnb was founded after renting sleeping bags in a crowded Washington DC, or how an inexpensive car became an idea after seeing a family on a two-wheeler. The reader can easily recognize innovation capital and habitus by recalling acknowledged innovators like Bill Gates, Steve Jobs and Elon Musk. Breathless media reportage about start-ups and founders adds to those entrepreneurs' innovation capital and habitus; conversely, the relatively lower capital and habitus give an institution less innovation power.

Innovation adoption is the most important and regrettably under-recognized factor. It is about how many consumers adopt the innovation *at a price level that creates a sustaining business model*. There is little point in getting adoption without profit sustainability or the converse. This is the dilemma that many Indian innovators face. Only when an innovator persuades the burgeoning, targeted millions of Indian consumers to adopt affordable water purifiers or feminine hygiene products at a sustainable margin can we sense the innovation adoption and impact. Several bottom-of-the-pyramid Indian innovations fail to develop high adoption levels or demonstrate the stamina of profitability.

Clever but low-impact ideas are a reality. Experiencing inadequate adoption for a clever idea happens all over the world. American innovator Frances Gabe died in obscurity the day after Christmas 2016 at the age of 101. She had devoted her lifetime to develop clever technologies that could reduce the drudgery of housecleaning. She designed the world's first and only 'self-cleaning home'. What a fabulous idea! In 1984, she got a patent for her invention and the self-cleaning home built by her had sixty-eight individual inventions. In her heyday, she acquired a fair amount of public profile as dictated by her innovation capital and habitus. Alas, the impact of the innovation was low because not enough people adopted the innovation. You may not have even heard of her!

The focus of our frenetic 'awards environment' needs to shift from clever ideas to innovation adoption. Innovators and judges of innovations should shift their attention to actual adoption through a sane business-model amongst the three factors of innovation capital, habitus and adoption.

Notes

Preface

1. Norman Doidge, *The Brain That Changes Itself: Stories of Personal Triumph from the Frontiers of Brain Science* (New Delhi: Penguin, 2007).

2. Rick Hanson, Hardwiring Happiness: *The New Brain Science of Contentment, Calm, and Confidence* (New York: Random House, 2013).

3. Pagan Kennedy, 'To Be a Genius, Think Like a 94-Year-Old', the *New York Times*, 7 April 2017, http://nyti.ms/2wcQ6nq.

4. Daniel Kahneman, *Thinking, Fast and Slow* (London: Allen Lane, 2007).

Introduction

1. Emma Green, 'Innovation: The History of a Buzzword', *Atlantic*, 20 June 2013, http://theatln.tc/2s3U0MX.

2. Tyler Cowen, *The Great Stagnation: How America Ate All The Low-Hanging Fruit of Modern History, Got Sick, and Will (Eventually) Feel Better* (New York: Penguin, 2011).

3. John Medina, *Brain Rules* (Seattle: Pear Press, 2008).

4. Yuval Noah Harari, *Sapiens* (New York: Penguin Random House, 2015).

5. Olivia Fox Cabane and Judah Pollack, *The Net and the Butterfly: The Art and Practice of Breakthrough Thinking* (Toronto: Penguin Random House, 2017).

6. Erik H. Erikson, *The Challenge of Youth* (New York: Doubleday Anchor, 1965)

I Fertilization: The Concept Gets Fertilized in the Brain

1. Norman Doidge, *The Brain That Changes Itself: Stories of Personal Triumph from the Frontiers of Brain Science* (London: Penguin, 2007).

2. Joanna Klein, 'Secrets of How the Brain Works', the *New York Times*, 17 February 2017.

3. Steven Johnson, 'Where Good Ideas Come From', TED Talks, September 2010.

4. Timothy Prestero, 'Design for People, Not Awards', TED Talks, August 2012.

5. Robert Douglas Friedel, Paul Israel and Bernard S. Finn, *Edison's Electric Light: Biography of an Invention* (New York: Rutgers University Press, 1987).

6. Andrew Hargadon, *How Breakthroughs Happen: The Surprising Truth About How Companies Innovate* (Boston: Harvard Business School Press, 2003).

7. Alexander Tsiaris, 'From Conception to Birth', TED Talks, November, 2011.

8. Adam Grant, 'How to Build a Culture of Originality', *Harvard Business Review*, March 2016.

II Birth: An Idea Is Born out of This Concept

1. Matt Ridley, 'When Ideas Have Sex', TED Talks, July 2010.

2. Steven Johnson, *Where Good Ideas Come From: The Natural History of Innovation* (London: Allen Lane, 2010); and also Mario Livio, *Brilliant Blunders: From Darwin to Einstein—Colossal Mistakes by Great Scientists That Changed Our Understanding of Life and the Universe* (New York: Simon & Schuster, 2013).

3. Philippa Neave, 'The Unexpected Challenges of a Country's First Election', TED Talks, October 2016.

4. Sandra Aamodt and Sam Wang, *Welcome to Your Brain: Why You Lose Your Car Keys but Never Forget How to Drive and Other Puzzles of Everyday Life* (New York: Bloomsbury, 2008).

5. Carol S. Dweck, Mindset: *The New Psychology of Success* (New York: Random House, 2006).

6. Louis Carter, 'Business Literature Archive', *strategy+business*, 28 October 2011.

7. Carol Sanford, *The Responsible Business: Reimagining Sustainability and Success* (San Francisco: Jossey-Bass, 2011).

8. Annie Murphy Paul, *Origins: How the Nine Months Before Our Birth Shape the Rest of Our Lives* (New York: Hay House, 2010).

III Infancy: A Prototype That Expresses the Idea

1. Richard Florida, *The Rise of the Creative Class: And How It's Transforming Work, Leisure, Community And Everyday Life* (New York: Basic Books, 2012).

2. M. Reznikoff, G. Domino et al., 'Creative Abilities in Identical and Fraternal Twins', *Behaviour Genetics 3*, no. 4 (December 1973): 365–77.

3. Jonah Lehrer 'The Itch of Curiosity', *Wired*, March 2010.

4. George Loewenstein, 'The Psychology of Curiosity: A Review and Reinterpretation', *Psychological Bulletin* 116, no. 1 (1994): 75–98.

5. Alan Deutschman, 'The Fabric of Creativity', *Fast Company* 89 (December 2004): 54–60.

6. Pico Iyer, 'The Art of Stillness', TED Talks, November 2014.

7. Sudha Shah, The King in Exile: *The Fall of the Royal Family of Burma* (New Delhi: HarperCollins, 2012).

8. Modupe Akinola, 'How Setbacks Spur Leaps Forward', Columbia Business School, 9 December 2016, http://bit.ly/2ll0LTz.

9. Sarah Parcak, 'Help Discover Ancient Ruins', TED Talks, January 2017.

10. Nayanjot Lahiri, *Finding Forgotten Cities: How The Indus Civilization Was Discovered* (New Delhi: Permanent Black, 2005); and Charles Allen, *Ashoka: The Search for India's Lost Emperor* (London: Hachette Digital, 2012).

11. David Burkus, *The Myths of Creativity: The Truth About How Innovative Companies and People Generate Great Ideas* (San Francisco: Josey-Bass, 2013).

IV Childhood: A Model Shaped out of This Prototype

1. Bernd Kriegesmann, Thomas Kley and Markus G. Schwering, 'Creative errors and heroic failures: capturing their innovative potential', *Journal of Business Strategy* 26, no. 3 (2005): 57–64.

2. Charles J. Pellerin, *How NASA Builds Teams: Mission Critical Soft Skills for Scientists, Engineers, and Project Teams* (New Jersey: Wiley, 2009).

3. Malcolm Gladwell, *Outliers: The Story of Success* (London: Allen Lane, 2008).

4. Steve Denning, 'Telling stories', *Harvard Business Review* (May 2004), http://bit.ly/1gN0s1e.

5. Joseph Campbell, *The Hero with a Thousand Faces* (Novato, California: New World Library, 2008).

V Adolescence: A Product Shaped from this Model

1. V.A. Shiva Ayyadurai, *The Email Revolution: How to Build Brands and Create Real Connections* (New York: Allworth Press, 2013).

2. Margalit Fox, 'Daniel Thompson, Whose Bagel Machine Altered the American Diet, Dies at 94', the *New York Times*, 21 September 2015, http://nyti.ms/2vLgAcB.

3. György Moldova, *BallPoint: A Tale of Genius and Grit, Perilous Times, and the Invention That Changed the Way We Write* (North Adams, Massachusetts: New Europe Books, 2012).

4. Steven Johnson, *Where Good Ideas Come From: The Natural History of Innovation* (London: Allen Lane, 2010)

VI Young Adult: The Product Competes to Grow

1. Michael S. Malone, *The Intel Trinity: How Robert Noyce, Gordon Moore and Andy Grove Built the World's Most Important Company* (New York: HarperCollins, 2014).

2. T. Thomas, *To Venture and to Build* (Kerala: Manorama Books, 2015).

3. Omkar Goswami, *Goras and Desis: Managing Agencies and the Making of Corporate India* (New Delhi: Penguin, 2016).

4. Bill Gross, 'The Single Biggest Reason why Start-ups Succeed', TED Talks, June 2015.

5. Giles Wilke, 'The Misplaced Hero Worship of Founders', the *Financial Times*, 5 June 2015.

6. Ashish K. Mishra, 'How an Indian e-commerce firm ran out of cash', *LiveMint*, 3 July 2015, http://bit.ly/1HAmwWZ.

7. R. Gopalakrishnan, *What the CEO Really Wants from You: The 4 As for Managerial Success* (New Delhi: HarperCollins, 2012).

8. Jose Guimon, *Promoting University–Industry Collaboration* in *Developing Countries*, Innovation Policy Platform, OECD and World Bank, 10.13140/RG.2.1.5176.8488.

9. Shail Kumar, *Building Golden India: How to Unleash India's Vast Potential and Transform Its Higher Education System. Now* (California: ONS Group Press, 2015).

10. H.C. Srivastava et al., *Pulse Production, Constraints and Opportunities: Proceedings of Symposium on Increasing Pulse Production in India,* Hindustan Lever Research Foundation (New Delhi: Oxford & IBH, 1984), XIX, 493.

11. R.A. Mashelkar, *Reinventing India* (Pune: Sahyadri Publications, 2011).

12. Julio A. Pertuzé, Edward S. Calder et al., 'Best Practices for Industry–University Collaboration', *MIT Sloan Management Review* 41, no. 4 (Summer 2010): 82–90.

13. Michael L. Tushman, 'Special Boundary Roles in the Innovation Process', *Administrative Science Quarterly* 22, no. 4 (December 1977): 587–605.

14. Lance Lee, David Magellan Horth and Chris Ernst, *Boundary Spanning in Action: Tactics for Transforming Today's Borders into Tomorrow's Frontiers*, Center for Creative Leadership, white paper, 2014, http://bit.ly/2v4tNvb.

15. Porus Munshi, *Making Breakthrough Innovation Happen: How 11 Indians Pulled off the Impossible* (New Delhi: HarperCollins, 2009).

16. Lynda Gratton, *Hot Spots: Why Some Teams, Workplaces and Organizations Buzz With Energy—And Others Don't* (San Francisco: Berrett-Koehler, 2007).

VII Maturity: Challenged to Change

1. Kiran Karnik, *Crooked Minds: Creating an Innovative Society* (New Delhi: Rupa Publications, 2017).

2. Sushanta Kumar Sen, *Corporate Sector Involvement in Sanitation*, Confederation of Indian Industry, 13 August 2013, http://bit.ly/2wSBnuT.

3. Giulia Enders, *Gut: The Inside Story of Our Body's Most Underrated Organ* (Brunswick, Victoria: Scribe Publications, 2015).

4. Sam Grobart, 'The Cult of the Squatty Potty', Bloomberg.com, 23 December 2015, https://bloom.bg/2fO8yLQ.

5. Jareen Imam, 'San Francisco starts using pee-proof paint to stop public urination', CNN.com, 25 July 2015, http://cnn.it/1fv6kLI.

6. www.sulabhinternational.org.

7. 'The Final Frontier', *The Economist*, 19 July 2014.

8. Jennifer Alsever, 'Turning waste into gold', *Fortune*, 12 June 2017, http://for.tn/2sgNwK4.

9. Y.S.P. Thorat and R. Gopalakrishnan, *What India Can Do Differently in Agriculture: Sarthak Krishi Yojana*, October 2015, http://bit.ly/2fO3nvD.

10. Malcolm Gladwell, *The Tipping Point: How Little Things Can Make a Big Difference* (Boston: Little Brown and Company, 2000).

11. T.N. Ninan, *The Turn of the Tortoise: The Challenge and Promise of India's Future* (New York: Penguin, 2015).

12. Leigh Gallagher, 'The Original Hospitality Disrupter', *Fortune*, 8 June 2017, http://for.tn/2uLu7n0.

13. 'How sham food became big business in Japan', *The Economist*, 8 June 2017, http://econ.st/2fNYPp8.

14. Pam Belluck, 'Chilly at Work? Office Formula Was Devised for Men', the *New York Times*, 3 August 2015, http://nyti.ms/2v49jmc.

15. Marlon Brando in *A Streetcar Named Desire*.

16. Yufen Chen, 'Impact of Foreign Direct Investment on Regional Innovation Capability: A Case of China', *Journal of Data Science* 5 (2007): 577–96.

17. Dan Senor and Saul Singer, *Start-up Nation: The Story of Israel's Economic Miracle* (New York: Hachette, 2009).

VIII Ageing: The Struggle for Relevance

1. Mariana Mazzucato, *The Entrepreneurial State: Debunking Public vs. Private Sector Myths* (London, New York: Anthem Press, 2013).

2. *Technology Strategy Board Report*, UK Department of Business, Innovation and Skills, May 2011.

3. Innovation and Commercialization, Princeton University, 1995.

4. Gopichand Katragadda, 'At the Bend in the River', *Tata Review* (January–March 2017): 101–03, http://bit.ly/2i7Sci3.

5. 'Invasion of the bottle snatchers', *The Economist*, 9 July 2016.

6. www.globalwellnessinstitute.org.

7. Brett H. Robinson, 'E-waste: An assessment of Global Production and Environmental Impacts', *Science of the Total Environment* 408 (2009): 183–91, http://bit.ly/2wiIXAT.

8. R. Gopalakrishnan, *A Comma in a Sentence: Extraordinary Change in an Ordinary Family over Six Generations* (New Delhi: Rupa Publications, 2016).

9. Ian Mortimer, *Human Race: Ten Centuries of Change on Earth* (London: Vintage Books, 2014).

10. Clayton M. Christensen, Scott D. Anthony and Erik A. Roth, *Seeing What's Next: Using the Theories of Innovation to Predict Industry Change* (Massachusetts: Harvard Business School Press, 2004).

11. Anab Jain, 'Why we need to imagine different futures', TED Talks, 19 June 2017, http://bit.ly/2rC9Jz2.

Epilogue

1. Refer to news.ul.edu>articles>2017/6.